Pass It On

Pass It On

What We Know . . .
What We Want You to Know!

BILLY GOLDFEDER AND FRIENDS

Fire Engineering

> **Disclaimer**
>
> The recommendations, advice, descriptions, and the methods in this book are presented solely for educational purposes. The author and publisher assume no liability whatsoever for any loss or damage that results from the use of any of the material in this book. Use of the material in this book is solely at the risk of the user.

Copyright© 2014 by
PennWell Corporation
1421 South Sheridan Road
Tulsa, Oklahoma 74112-6600 USA

800.752.9764
+1.918.831.9421
sales@pennwell.com
www.FireEngineeringBooks.com
www.pennwellbooks.com
www.pennwell.com

Marketing Manager: Amanda Brumby
National Account Manager: Cindy J. Huse

Director: Mary McGee
Managing Editor: Marla Patterson
Production Manager: Sheila Brock
Production Editor: Tony Quinn
Cover Designer: Paul Combs

Library of Congress Cataloging-in-Publication Data

Pass it on : what we know, what we want you to know / Billy Goldfeder, editor.
    pages cm
 ISBN 978-1-59370-319-6
1. Fire fighters--Anecdotes. 2. First responders--Anecdotes. I. Goldfeder, Billy.
  HD8039.F5P37 2013
  363.37092'2--dc23
                        2013033086

All rights reserved. No part of this book may be reproduced, stored in a retrieval system, or transcribed in any form or by any means, electronic or mechanical, including photocopying and recording, without the prior written permission of the publisher.

Printed in the United States of America

4 5 6   18 17 16 15 14

CONTENTS

Read This First: Introduction by Billy Goldfeder 1

John Alston
The Seven Cs 17

Joseph J. Apuzzio
The Importance of the Support of the Brotherhood of Firefighters
to the Family after a Line-of-Duty Death................................. 23

Steve Austin
What I Have Learned... 27

Anthony Avillo
The Failure of Imagination and the Unintended Consequences 31

Douglas Barry
Fire Service Leadership 35

Marc Bashoor
Leading Forward—A Fire Chief's Expectation 39

Phil Bird
Today's Fire Officer... 43

Rebecca Boutin
Never Stop Training.. 47

Garry Briese and Oren Bersagel-Briese
The Badge, Family, and Success 51

Alan Brunacini
Equipment #1489 55

Alan Brunacini
Timeless Tactical Truths 57

John M. Buckman III
Creating a Survival Mentality... 61

Dr. Harry Carter
Avoiding the New Boss Blues 67

Sal Cassano
A Strengthened Sense of Purpose 73

James Clack
What Happened to Us? 83

Burton A. Clark
What You Do Wearing the Helmet Is Only as Good as What You Put
in Your Mind 87

Burton A. Clark
First Fire—Important Lessons 89

Hank Clemmensen
Acceptable Risks 93

Kelvin J. Cochran
Transformational Followers 97

Allyson Coglianese
Lt. Edmond P. Coglianese, Chicago Fire Department, Engine 98,
Last Call: January 26,1986 103

Eileen Coglianese
How We Honor Our Fallen 106

John "Skip" Coleman
My Advice to the Ranks 109

Ronny J. Coleman
The Bondi Story 113

Paul Combs
No Place for Bullies 117

Dennis Compton
Being a Firefighter Is a Privilege to Be Cherished 121

David "Chip" Comstock, Jr.
Caring 125

Glenn P. Corbett
Know Your History! 135

Dave Daniels
It Is Okay to Keep Studying 139

Peter Demontreux
Things I've Learned 143

Ty Dickerson
The Role of the Company Officer. 147

Dave Dodson
What You Don't Know Can Kill You 153

Chuck Downey
Communications 160

Joseph Downey
A New York Legacy 163

Michael M. Dugan
Don't Be a Wheelbarrow 169

Glenn A. Gaines
The Core of the Matter in Fire Service Excellence 173

Gordon Graham
Predictable Is Preventable 178

Bill Gustin
Wear Your Chinstrap. 183

Alexander Hagan
The Floor Above 187

Bobby Halton
Have the Courage of Your Convictions and Humility of Our Mission. . . . 191

Joanne Hayes-White
A Rare Opportunity. 197

Cathy Hedrick
Kenneth M. Hedrick, Morningside, Maryland, Volunteer Fire
Department, 1992. 201

Cheryl Horvath
What I've Learned 205

Otto Huber
Leave It Better than You Found It 209

Ron Kanterman
Leadership 211

Brian Kazmierzak
Just When You Think You Have Your Plan Figured Out 217

Tony Kelleher
Mentorship and Motivation in the Fire Service...................... 221

Pat Kenny
Learning a Lesson from Sean about Superman....................... 227

Stephen Kerber
Understanding and Partnerships Save Lives 233

Rhoda Mae Kerr
What I've Learned 237

Scott D. Kerwood
This Isn't Your Fire Chief's Fire Service Anymore! Stop Sniveling
about Education 241

Bob Khan
The Best Job in the World 245

Edward Kilduff
Get Squared Away... 249

William D. Killen
Overcoming Peer Pressure in the Fire Service....................... 253

Danny Krushinski
What You Don't Know Will Make the Difference between Life
and Death.. 259

Rick Lasky
The Firefighter's Time Machine 265

Edward Mann
Simple Truths, Hard Lessons 269

Richard Marinucci
What I Have Learned—Thoughts and Advice....................... 273

Ray McCormack
A Culture of Extinguishment.................................... 277

Bruce J. Moeller, PhD
Think!.. 281

Frank Montagna
Training Yourself... 285

Tom Mulcrone
A Sacred Task.. 289

Jim Murtagh
What Can You Do?... 293

Gregory G. Noll
What I've Learned.. 299

Denis Onieal
A Range of Thoughts from a Very Diverse Career............... 303

Steve Pegram
The Smartest Guys in the Room................................ 309

Joseph W. Pfeifer
Doing Ordinary Things . . . So Others Might Live............. 315

Shane Ray
Do It Different . . . Don't Risk It All!..................... 321

Jay Reardon
Promising People... 325

Frank Ricci
The Mission, the Men, and Me................................. 329

J. Gordon Routley
The Good Old Days Weren't All That Great—Especially
for Firefighters... 335

Dennis L. Rubin
The Best Discipline Is Self-Discipline....................... 341

John Salka
Keep Reading, Studying, and Learning......................... 347

José A. Santiago
The Value of Leadership...................................... 349

Selena Schmidt
A Mother's Legacy 353

William Shouldis
#1 Meridian Plaza—A Story of Tragedy and Change: A Night
of "High-Rise Horror" 357

Ron Siarnicki
No One Goes Home . . . Without Being Checked Out if They Become
Ill at an Incident Scene. 361

James P. Smith
What I Learned in Life and What the Fire Service Taught 365

James P. Smith
Firefighting and Cancer. 367

James P. Smith
Effective Fireground Operations 369

Ronald R. Spadafora
Risk vs. Reward: 7 World Trade Center 373

Dave Statter
Soon, the Most Trusted Source in News May Be the Local
Fire Department 379

Phil Stittleburg
A Thousand Thank Yous 383

John B. Tippett, Jr.
Keep Your Head When Others Are Losing Theirs 387

Matt Tobia
Service above Self 391

Bruce H. Varner
45 Years and Still Learning. 395

Curt Varone
Give Me Five Minutes Kid, Sit Down and Listen to What I Want You
to Know from My View of the Fire World 399

Curt Varone
Reflections on The Station Night Club Fire 401

Colleen Walz
It's Not a Cliché... 409

Bill Webb
Politics Is Local: You Can Make a Difference for Your
Fire Department... 415

Mike Wilbur
Six Lessons for Success... 419

Janet Wilmoth
Five by Five... 423

Arlene Zang
Brother, Where Art Thou?... 427

FF/Paramedic Brian Goldfeder and FF/EMT Dave Stacy . . .
Some Perspective from Some Improved Chips off Some Relatively
Old Blocks.. 431

Anonymous
Advice for the Brand New Firefighter . . . Well, Not Really a Firefighter.
Not Yet... 437

Read This First

INTRODUCTION BY BILLY GOLDFEDER

I have been a lover of fire department–related books since I was a little kid. My parents told me that I had really little focus on anything else. My school grades *always* proved that. My firefighting interest has been there ever since I can remember, and it's true—I have very little interest in much else. My focus has always been narrow and limited to being a firefighter, related friends, a drummer, and (longer term) being a dad (and who knew, being a *Poppie* would become part of it). That was pretty much it.

My first memories started growing up on Long Island, and in particular the Manhasset-Lakeville Fire Department (M-LFD); the Great Neck Vigilant Engine and Hook & Ladder Company; and the Alert Engine, Hook, Ladder, and Hose Company No. 1. Those three departments protected the Great Neck peninsula that I grew up on with M-LFD protecting the largest of the areas as their area spread further south and east. I would often get to go with my mom, dad, and sister into New York City, and there, of course, I would see the FDNY. FDNY. That was cool. *The* city. Many of the firefighters in my hometown were also FDNY firefighters, and that was always a very cool connection. My goal was to get onto the FDNY, but there was no Lasik surgery back then, so that took care of that. Things turned out pretty well anyway. Another goal was to get on M-LFD, and that did work out well. To this day, my pride of being a member of M-LFD has never waned. It is one of the finest and most professionally led volunteer fire departments anywhere—seriously. As a very little kid, my earliest memories of anything to do with firefighting started with M-LFD. Hearing the firehouse's outdoor sirens and horns blast, summoning the members, was a sound many of us grew up with. The sounds stuck in my head. Forever.

And speaking of forever, you'd be lucky if, like me, in your career you are privileged to work with good firemen such as Chief Pat McGrath, Captain John Brown, Chief Eddie Bennett (RIP), FF George Lucas, Captain Bobby Scalcione, Lt. Herbie Koota, FF/Medic Jon Orens, FF Eddie Sherman, Lt. Russ Randolph, FF Mike Protitch, the Pinellos, the Bernatovichs, the Costa and Hicks Brothers—a few of the "original" squad members from M-LFD

Company 3, 1970s. Real good firemen—and good guys—who saved countless lives and helped create memories that will last forever.

From M-LFD to Broward and East Manatee, Florida, to Loudoun County and here in Ohio, I've been very fortunate to work with some outstanding women and men who demonstrate their dedication to the job each and every day. I'm not talking about empty suits (we've had more than our share of those) but good, proven firefighters and medics who demonstrate their selfless dedication, regardless of the obstacles day in and day out.

In discussions, we so often ask each other what got us started in this. For some, it's as simple as needing a job, taking a test, and getting on—if you passed. Others had always wanted to be firefighters. I fell into the latter category. So what memories do I have? Tons. But let me share one with you.

Growing up in New York, my folks had a guy come in annually to strip and wax the basement floor. His name was Danny, and he was a firefighter in Commack, Long Island. Some knew him as Dino. He would bring his large, swirling, motorized floor waxing machine and run it though our basement. He would also let me sit on top of the machine as he ran it on the floor—as I held on tight! That was great! There were no Occupational Health and Safety Administration (OSHA) rules back then. Sort of like riding tailboard. So one day, Danny was taking the machine back out to his woody-style station wagon in our driveway, and as he lifted it in, I saw in the back of his car a black leather helmet, a black coat with small silver reflective stripes, and rubber pull-up boots. I knew what they were, but I didn't know why *he* had them. He went on to explain to me that he was a volunteer fireman (that's what they were called back then, *relax*) and when he was home, he was "on duty" as a fireman. Whoa. Seriously. *You* are a fireman. That was beyond great.

After that, each time Danny would come to our house, he would bring me copies of fire magazines such as *Fire Engineering, Volunteer Firefighter, Fire Command,* and the like. What really stands out, though, is that he also brought me training manuals from Suffolk County's fire training program. Worn out booklets on Firemanship 1, House Firefighting, and similar subjects that were, to me, like winning the lottery! I was in heaven and really felt that instead of Danny coming once a year to wax the floor, I felt strongly that Danny should be there doing it weekly. The more he could bring me, the more I would look at, read, and study. Fast forward to the late 60s, early 70s, and he continued to never forget to bring me "the good stuff," and that action had a profound impact on me, even more growing my interest in firefighting. He passed it on.

I started "officially" firefighting in 1973 and have loads of great stories about good, bad, and related stuff, but the focus of this book is to provide you with lots of ideas, suggestions, and advise mostly from firefighters, fire officers, and chiefs, as well as family members and friends of our profession. I'll save most of those other stories for another time.

Several years ago it was suggested that I write a book; however, I felt that it was already a pretty crowded field. I definitely enjoy writing for the magazines as well as what Gordon Graham, our team, and I do in attempting to covertly control the fire media, through www.FireFighterCloseCalls.com as well as the newsletter *The Secret List*. But as far as writing a book? That sounded like a lot of work, and I really didn't have an interest to write one on subjects such as survival, safety, tactics, operations, command, and so on—there are plenty of excellent ones out there, written by *the best of the best*. So I thought about what I might be able to offer if I were to "do" a book.

What I am giving you through this book are the thoughts, ideas, and advice of the many people I know, or are related to, in the fire business. With 40 years as a firefighter, I have been very, *very* fortunate to have become friends with a lot of people both in local and national fire and EMS. People that have so much to pass on. And that is what got me started on this book. The names of the contributors to this book were not just picked out of a hat, but are people I know who have messages for other firefighters, probies, officers, future officers, and so on. I had to whittle down the number of contributors and who they are. I wanted to keep it diverse so that we could pass on advice and thoughts to as many firefighters as possible. Deciding on the contributors for this book wasn't easy, but it was not about any of them—or me. The focus of this book is in two parts: passing along what we know to benefit other firefighters and raising some much-needed funding for two important charities. There are many folks on deck for what might be volume 2 of this book, most of whom don't even know it yet.

Additionally, making money is not the object of this book, and we have worked to keep it very affordable. Not unlike when I did the Pennwell/Fire Engineering video years ago, *And The Beat Goes On*, it is important to give back. I don't think we give back enough, both in life as well as firefighters. Sometimes we find ourselves in this *what-the-world-owes-me* fog, and we need to be snapped out of it. I'm not sure the world owes us anything.

Not a single contributor in this book is being paid anything—nor am I. One hundred percent of the writer royalties and fees will be shared 50/50 between the **Chief Ray Downey Scholarship Charity Fund** and the **National Fallen Firefighters Foundation.** My logic is that most of what we are sharing with you in this book isn't *owned* by us. The advice and suggestions being passed on to you are based upon experiences that each of us have been through. Some experiences are great, some are not very flashy but important, and some are absolutely horrible. It runs the entire gamut. Some of the experiences the writers are glad to recall and remember, and others are based upon the worst day ever in the lives of some who contributed to this book. They are all sharing, passing it on to you, and giving back. Read every one of them. *Trust me.*

When I reached out to the final list of contributors, I told them that I simply wanted them to share their advice to any group or groups within the fire service. It was totally up to them; the field was wide open. I did no editing; that was done by the PennWell book folks, all of whom have no fire service experience. That created an environment where what the writers wrote is what you are reading with just grammatical editing. As you can also see, when we reached out to Paul Combs to do the cover art, they gave him total freedom as well. Now looking at how the cover turned out, in the future I may rethink "creative freedom!" I also, with some independent advice, picked a very diverse group of writers. As you will find out, they definitely do not all fully agree on the same things tactically, operationally, or generally. But what they all do agree upon is that they have a message and advice to pass on to you. It is up to you to determine how you use it, and when.

THANKS

My sincere thanks to the contributors to this book. I won't list them all here; you will see their names and stories in the table of contents. In one way or another, I have interacted with all these people, and it is due to that interaction that they have been gracious enough to contribute. If I bring anything to the table, I feel pretty strongly about maintaining relationships with good people and passing it on between us in person, by phone, and in writing. Each one of the contributors understood that their mission in this book project was to "pass it on" so that others such as you can gain from their experiences and advice. The contributors range from fire commissioners and chiefs in America's largest departments to leaders from some of the smallest departments. Career, volunteer, union, nonunion—they are all included. Some of the contributors are not firefighters but have so much to offer, such as some of Americas best risk attorneys to the immediate family members of firefighters who were killed in the line of duty. Some have been company officers most of their careers, and a few are relatively young firefighters to help bridge any generation gap concerns. Some of the writers I have known for decades, and some I have known for just a few years. They range from high-school-diploma holders to doctors and PhDs, from those who have demonstrated superhuman bravery in saving lives to those who teach us how to do so. Some have saved lives by directing companies and talking on the radio, some in the classroom, and some by crawling down hallways. I have tried to provide you with an opportunity to meet and hear from some of the best in our business, people who you might not be able to gain knowledge from personally, as intimately, in one large, diverse source.

I titled each one of the contributors' sections/chapters based upon how I felt about them, their message, and our relationships. Some of the titles directly go into what they wrote about; some don't. Furthermore, I wrote the introduction to each of them, so that you would have a little idea of what I think about them and the good they bring to the table. For some of them, I mention awards they have won; for others, I don't. For some, I mention books they have written, and some I don't. The rhyme and reason was pretty much the manner in which I would introduce them to you personally, to gain your attention, so that you will read, listen, and learn from their experiences. We are *really* fortunate. Finally, a very special thanks to those who helped edit this book, especially to my friends Mollie Pegram and Dave Stacy for their painstaking work. The hundreds of hours they donated to this project are priceless. Naturally, if there are any errors or omissions, blame them.

Although you understand I love being a firefighter, I also have a few other love priorities. I love music. Some of you can tell stories about fires—and music. Note that I didn't ask you to write for this book. Not yet—maybe the next one. I've been a drummer since the fourth grade and always grew up around music. I happily blame that on my mom, because growing up in New York there was always a radio tuned to WMCA or WABC (AM). I continued that with my kids since day one and with my grandkids today. It's good stuff.

If you have ever heard me speak, or have even just chatted with me, the love for my family is always part of it. Actually, I force them on you. That's how it works with me and our family circus. My daughter Amy (an elementary public school teacher) and her husband Thomas, their two sons, our "way cool" grandsons Henry and Camden; my daughter Dani (special education school teacher), her husband Matt, and their beautiful little baby girl, our granddaughter Vada; my (firefighter/paramedic) son Brian, his wife Lindsey, and their beautiful little baby girl, our granddaughter Harper; they are all *within every second* of my life and every beat of my heart. Seriously. Thanks to my sons James and Sean, who entered our lives (with their mom Teri) a bit late, but provide us with smiling perspective as they work their way through this world of ours with their unique views and equally caring hearts. The closeness I have to all of these kids, all of them, is beyond precious to me—and I am beyond fortunate.

My dad (Sam) passed away a few years ago and is missed every day. He was a man who directly (and initially indirectly) inspired me to do what I do. A member of the greatest generation, he was a child immigrant to the United States, became a decorated World War II veteran who then went on to be a fiercely proud and patriotic businessman, son, brother, husband, dad and "Pop-Pop." Real American Dream stuff. My sister Sue has had a huge ladder to climb most of her life and has done so well, regardless of the obstacles thrown. To my mom, Mrs. G, Joyce. Her energy, great sense of humor, positive and caring attitude, along with her great outlook on life is beyond inspiring. There are few great grandmas who are e-mailing and IMing to stay in touch with their grandkids and great-grandkids every day. Her focus on health and acceptance of progress balanced with her back-in-the-day down-to-earth sense and outlook *always* clears things up. For example, "the weather used to be a lot better and more predictable until they started shooting all those damn rocket ships up into outer space—and now look at it, what a mess." *You can't argue with that.* And speaking of not arguing, thanks to my wife Teri. My personal rapid intervention team of one, her balance, careful and cautious "size up" view on life has made a huge difference to me, and to all of our kids. Her ability to match humor, wit, and words with me has been eye opening and makes it where we rarely want to let go of one another. A firefighter/medic, her very visible *way beyond normal* deep caring and deeper love for our family—and especially me, is her standard operating procedure. As you read this book, you will understand why I consider myself a very blessed and fortunate (*young*) man.

Enjoy.

SOME THOUGHTS
I WANT TO PASS ON TO YOU

Size-up in more ways than one (before, on arrival, and during the fire)

As a firefighter, there is no one way of doing most things. It generally depends on the conditions, although these days we seem to be looking for *the* answer. Quit looking; there is none. It all depends on what I refer to as the three-way size-up.

What do we have?

What do we want to do?

What are our resources?

Recently, there have been some eye-opening scientific studies led by veteran firefighters (several who have contributed to this book) related to operating at fires, and in particular, single and multifamily dwellings. Those fires are the majority of structural fires that all of us respond to, generally (but not always) with significant personal risk. Risk to those who own the homes and risk to us, attempting to save people and their stuff.

Regardless of those who say there are no fires anymore, there are fires. People are still being killed in those fires, and firefighters are challenged to do more with less. One of the goals of the many studies is to help us best understand how we can reduce the bad stuff from happening. Stop the fire. Stop the victims from getting hurt or killed, and sometimes stop us from getting hurt or killed.

So what's the best way to operate?

Well, until the careless big-buck homebuilder associations stop their 50-shades-of-shame political love life with some state and city hall dwellers, residential fire sprinklers will sit on the sidelines as the answer to "Why did the house have to burn down?" and "Why did they have to die in that fire?" Fire sprinklers will essentially solve the fire loss and fire death problem, when the elected officials want it to. In other words, everything won't be sprinklered anytime soon.

So what is the *correct way* to operate at a single-family dwelling? What about a multifamily dwelling? Is it direct attack? Is it interior attack? Should I use a PPV fan? Should we vent the roof? Is this a transitional attack? The statement "We are going with exterior attack from now on at all fires" is just as wrong as the statement "We are always an aggressive interior attack department."

The answer?

Show us the fire. Plain and simple. There is no *one way* to operate at a fire if you have the resources to apply options. That means if you have two firefighters responding to all fires, you actually have few choices. You are an exterior-attack, deck-gunning, water-loving department. Few options. However, if you are like most departments, you probably have several companies responding with staffing, which gives you some variable options to operate. And that's the issue. You have to size it up before you know what you are going to do.

Size it up before the fire. Size it up on arrival.

Size it up before the fire. Size up your response area, and plan ahead. How many tasks you want to perform equals how many firefighters you will need. What are the buildings and stuff you protect? What turns out on your first alarm assignment? Does it match what's being reported on fire? Will you have enough firefighters arriving in time so the tasks can be performed simultaneously . . . or eventually? One important goal is to not run out of a house before the predictably needed troops arrive. There are few fires that should surprise us when we arrive. To avoid being surprised, get out and look around; plan for the fire before the fire that will eventually come in. Responding to and arriving at a house fire in your first-due area should not be like arriving on a blind date. All the buildings in your first-due area are just sitting there, waiting for you to plan on the fire—well before there is a fire.

When arriving at the fire, you size it up, do a 360, determine whatcha got. Smoke? Fire? Occupants? You know what to do. Then you immediately determine what you want to do about it . . . and then you make sure you have the resources (firefighters and equipment) to do what you wanna do. Got the resources? Wonderful. You don't? Strike more alarms and then you'll probably want to start flowing a lot of water. Depending. Pretty standard stuff.

The fact is that with all these studies and the science that comes from them is the fact that we have even more options. More information. More training. More options to stop the fire and do all the other stuff we want to do. Hit the fire before going in? Yep. Enter from the unburned, and hit the fire? Yes. Aggressive interior attack? Sure. VEIS? Si. Blitz attack? Oui. Show us the fire.

The is no *one way* to do much in what we do; it's all about knowing what we have, deciding what we want to do or get done, and recognizing what resources we have in order to do *that*. The one area where we seem to get ourselves in trouble lately is the *wanting* to do without the resources to do. So what do you do? That question needs to be answered before the fire. Way before the fire.

There are some common denominators to all of the contributors in this book (firefighters should be well-led, well-trained, and not unnecessarily get hurt or killed, for example), but they also offer you a wide range of *options*. Just as I described earlier about size-up (before, on arrival, and during operations), based upon your situation, determine what you would like to accomplish and what resources you have to get that done, then apply their advice, thoughts, ideas, and philosophies in the same manner—*depending upon the situation*.

Keep in mind that just because you have done things the same way for years, like operating at a fire, there are sometimes different (or better) ways of doing things. Of course, right now you are thinking tactically. Forget that for a minute. This is more than just tactical. Consider the ideas that the contributors are providing you with as new ideas. In other words, rethink the ones you are most uncomfortable with. Give it a chance. Some of these folks have been in this business for a long, *long* time, and each of their ideas is presented to give you the chance to new-think some things, and rethink others. Sort of like accepting new technology. Sure, the old way worked pretty well, but son of a gun, this new way may work even better. Size it up.

We may not have a chance to size up before a blind date, but we can almost always size up before, upon arrival, and during fire.

Size up your people

We are quick to say that our people are our greatest resource, but how much time do we spend sizing them up? Strengths? Weaknesses? Needs? Skills? As a company officer, understand that they are not all the same, even though they have been created equal. Can you count on every member of your company to be experts on every task and assignment your company may be assigned? Size them up, and after you do that, determine what the needs are to get them all at the needed skill level.

We had a fire

Actually, you didn't have anything. Some poor individual had a fire, dialed 9-1-1, and hoped like hell that you actually had an idea about what to do with *their* fire when you arrived. In other words, it was not *our* fire, it was theirs. Now, take that and convert it into *your* fire. Let's just pretend that you have personally called for *your* fire department because you have an emergency. Now, what do you want done? How well trained should the personnel be? What leadership would you want commanding your fire? Do you want everyone even remotely connected to your fire (dispatchers, firefighters, officers, and chiefs) drug- and booze-free? In decent shape? No criminal backgrounds? What kind of staffing would you like? What kind of relationships would you like the firefighters to have with each other? Would you like them to understand the value and importance of water at a fire? Would you like the hydrants to be in perfect working order? What kind of tactics would you like performed if your loved ones were inside? How quickly would you like searches completed? Would you like apparatus checks done *before* the fire so that the stuff all works as designed when they arrive? If it were your fire, what would you like done?

I have found in my career that a good template is this: If something is good for the public and is good for the firefighters, we generally don't go wrong. And don't confuse this with being good for the firefighters individually, but more as what is best for them so they can best serve those dialing 9-1-1 who needed our help five minutes ago.

We are the best firefighters—ever

Sometimes our egos get in the way. As Bruno says, egos eat brains. In other words, sometimes we think we are better firefighters than those other jerks down the street. Sometimes that's correct, but you don't need to put it on a patch, write it on your apparatus, or run your mouth. Just show up and do your job following training, training, and more training. If the job you do is good, the reputation will follow. It is that simple. You and your company's actions, not words, will speak for themselves.

No fear? Be smart and fearful

A close friend of mine, instructor and veteran Chief Fire Officer Timmy Delehanty, once ripped a firefighter for having a No Fear sticker on his helmet. He wasn't concerned about the sticker, but about the mentality of that phrase. He asked the firefighter what it meant, and the young man looked at him as if he were crazy. The firefighter wasn't sure what to say. Timmy proceeded to make it real clear that the firefighter better be smart and *fearful,* and that anyone arriving at a fire better have some fear coupled with the required training in order to know what to do based upon the circumstances. Timmy is widely known as a good guy and quite a character, but when it comes to taking care of his people, there is no higher priority to him than making sure that training and experience are critical to fear management. We respond to bad stuff; a little managed fear based upon your understanding of what is needed to be a good firefighter will help.

Loyalty: Everything doesn't suck

Years ago, when I was a very young firefighter, the following advice was posted in a boss's office. I always remembered it; it stuck in my mind. This is good for any of us at all ranks, from firefighter to chief, girls and boys. While loyalty can definitely be challenging at times, and providing (and welcoming) input to change and improve is absolutely critical, this piece is a good reminder.

> If you work for a man, in heaven's name work for him. If he pays you wages which supply you bread and butter, work for him; speak well of him; stand by him, and stand by the institution he represents. If put to a pinch, an ounce of loyalty is worth a pound of cleverness. If you must vilify, condemn, and eternally disparage, resign your position, and when you are outside, damn to your heart's content, but as long as you are part of the institution do not condemn it. If you do that, you

are loosening the tendrils that are holding you to the institution, and at the first high wind that comes along, you will be uprooted and blown away, and will probably never know the reason why.

—Elbert Hubbard

To be clear, it's not always easy to be loyal. I get that. In some areas there are even so-called *fire* officers who have never served as firefighters. Rare; dangerous; insanity; but it has been allowed to happen by the clueless. However, generally those who outrank you *have* done what you are now doing. Consider that. Also consider that some of those same folks made the very *excellent* decision to hire you. Think about that when questioning their decision-making abilities.

We are here to serve. That includes the occasional exciting and challenging calls as well as the day-to-day calls. This isn't *Backdraft*, *Chicago Fire*, *Rescue Me*, or *Ladder 49*. That's Hollywood; this is reality. We can sometimes find ourselves getting caught up in what is referred to as the BS calls. *This is not about you.* Someone called 9-1-1. Did they need 9-1-1 from our perspective? Probably not. But for whatever reason, they called and you were dispatched. Yes. So often it is a waste of resources and yadda yadda, but you are in the "sometimes the calls don't always make me feel like a firefighter, darn it" business, so get past it. Granny pooped her pants . . . why were *you* called? Apparently no one else gives a damn about Granny, so now you are in the "we give a damn about Granny because her children are too busy with their own lives to care about her" business. Who else can she call? Be careful and remind yourself (and your subordinates) that we are here to serve, not self-serve. Want to feel more like a firefighter? Train. Then train more. If it still sucks, maybe this simply isn't the career (paid or volunteer) for you.

Bunker gear does not make a firefighter

Hopefully like you, I love being a firefighter. My enthusiasm is reportedly rather obvious. Apparently I have, at times, been unaware that not everyone in the business feels the same way I do. In 1982, just prior to becoming a chief officer, a friend of mine, a veteran firefighter who gets a kick out of my enthusiasm for the job, reminded me that not everyone feels the same way about the fire service as I do. He reminded me to remember that as I went from company officer to chief officer. Some are just here to do a job. Some are known as *square rooters*, people who calculate how they can gain as much as possible from the job by giving as little as possible. It's all about them. Know the type? Some are here to take up space. Some have other things going on in their lives than just the fire department. My friend *had* to be wrong. He was not. I get that there is much more than the fire department, and we definitely have to balance our priorities. In this case, I am specifically talking about the square rooters, the ones who give little and take a lot, the ones who are consciously malicious and for some reason were brought onto the fire department anyway—sometimes brought on by the very people they end up square rooting. *How about that?* As I say a few times in this book, I have been insanely fortunate and have had (and continue to have) a great career. As part of the playing field, I have had to deal with a few professional square rooters in my career, above and below my rank, and always took them on head-on. That has bitten me a few times, which can happen when chasing rats. I chased them without any regret because you are either into this job or you are not. Karma has done well in taking good care of the square-rooters. Just because someone wears bunker gear, that does not make that person a firefighter, a brother, sister, or part of the crew. Time will tell based upon actions. Avoid the square rooters, the rats, and those who, based upon a fair size up by you and others, have no business being here—and let the right people know it.

They do not think the way we do

The folks in city hall generally think we are really weird. Our culture is a curiosity to many of them. In so many cases they don't feel we should be getting what we get. Who the hell do we think we are anyway? I recently spoke with a chief from a department who suffered a multiple-firefighter line-of-duty death. He stated to me that while they are rebuilding and recovering from their worst fire, the city hall folks and other city departments are about fed up with the fire department getting everything. Yep. That's how it works. Who fixes that? Generally, that's the fire chief's job, to make sure the department is well represented to the city hall dwellers and the elected officials. To get the necessary resources and to best represent the facts to those folks. While we can kick, scream, and yell about people dying, the city hall folks rarely prioritize the way we do—and they even more rarely share the emotion we share, because they have never done what we do. The key seems to be a fire chief, association president, local president, or commissioner who can speak their language, the language of numbers and statistics proving what we know as fact. Take the emotion out, put the numbers in, don't be politically naive, and now you are starting to speak their language. Have trouble doing that? You may be better off being a deputy chief than being the chief of department.

There are choices in fire protection

We want it all *now*. When the tones go off, those focused on what's best for the public and us, want it all—right now—to arrive, force entry, search, vent, get water on the fire . . . you know the routine. Some communities simply cannot afford what we know they need. In some cases, that can be dealt with through collaborative relationships, automatic mutual aid, group purchasing, and similar mutual programs. The reality, however, is that what we think is best may not be what the chief thinks is best and may definitely not be what the city manager thinks is best. The latter is focused on the cost. *But what about the cost of a human life?* Don't bring that up. See earlier in this chapter about our thinking . . . and theirs. Elected officials and others in city hall seem to do best when given choices. The following model seems to work well.

A. The best for the citizens and the firefighters
B. A good plan for the citizens and firefighters
C. A basic plan for the citizens and firefighters

I liken this to your personal insurance. It is all based upon the risk you are willing to take, what you can afford, and finding that balance. For example, you can cut your car insurance and hope you don't have a claim, but the difference is that when you cut your insurance, you made that decision. In the fire service, in so many cases, the public fire protection has been cut, but the public, the *customer*, is clueless as to the impact that may have. They are fine until they dial 9-1-1 to make a claim and find out the reality of their coverage. When you cut your personal insurance, your agent will show you the impact that it will have if you make a claim. When some fire chiefs cut the budget, they do not share the realities of those cuts. An example is telling the public and the elected officials that browning out, or cutting fire station staffing, will not have an impact on the service delivery. Stop. *Stop the BS.* If cuts have to be proposed, propose them honestly and fairly, but make it clear through facts and numbers what the realities of those cuts will be and what changes your personnel will be forced to make based upon those cuts.

I don't like it, but it is the reality. If I were king, the parks would be shut down before we become unable to provide the best in fire/rescue/EMS protection. All hail.

The solution is to provide choices based upon fact and reality. Evaluate your community; determine the needed fire flows based upon the areas, building types, and construction; and set some goals, such as how many tasks you want performed in what period of time, how quick you want companies arriving, and how many firefighters are needed. If we are going to educate the city hall folks, I have learned that it takes less emotion and more speaking their language, and I have kicked and screamed all the way.

Nice wheels

Each month in *FireRescue* magazine, Chief Bob Vaccaro writes about fire and rescue apparatus, the importance of understanding what you are specifying, and the job that apparatus is expected to do. In *Firehouse* magazine, Lt. Mike Wilbur and Tom Shand provide similar critical advice on being an architect for your apparatus. Chief Bill Peters also provide similar expert advice in *Fire Engineering* magazine based upon his decades of experience. What do they all have in common? Expertise and experience that most of us do not have for specifying apparatus costing, on average a half million dollars and up. The advice here is simple—learn from those who know. Read the articles, and follow their advice. Maybe your department would do well by working with neighboring departments and an expert's help on purchasing a standard pumper. You are spending someone else's half million for that new apparatus; pretend it's coming out of your pocket, make sure it is what's needed, and that the specifications reflect it.

Nice clothes

We spend years designing fire apparatus so that it is just right for our specific needs. Of course, I'm not sure that the fire in Town A is so different from the one in Town B that each town needs a totally different specified fire apparatus, but that's a discussion for another day. My comparison here is between the time and interest we take in apparatus, and the time and interest we take in specifying our bunker gear. All bunker gear is not alike, and just like fire apparatus it is all based upon the specifications. Some say that they don't need the same heavy-duty gear specification as say, another town, because they don't have as many fires. That's parachute mentality. Jump from a plane using a minimally specified low-bid parachute? Or would you rather have one that's the best of the best? It's what is going to manage what happens between you in the air and your arrival on the ground. Same with your gear. Specify your gear as you do the apparatus. Don't accept low-bid, minimal specifications for what may be the only thing between you and the fire.

Go there

You need to go the National Fire Academy. This advice is for officers and those who want to become officers—career or volunteer. The cost is free; your tax dollars already paid for it. The opportunity to learn about subjects that apply to the job you do, or will do, is priceless. The opportunity to go Emmitsburg is twofold: the training you will get and the people you will meet. In every case you will come back with a much more in-depth perspective on your own department and yourself. Go to their website at http://www.usfa.fema.gov/nfa/, look at the course catalog, and make a plan.

So, that's just a few thoughts I wanted to pass on to you, from me. Now, enjoy much, much more from some wonderful people in our business, who simply just want you to do good.

The youthful author of this book doing some old-school training "back in the day" (1977) as a member of the Manhasset-Lakeville FD, Long Island, NY. Along with Billy is Firefighter Jon Orens (now Dr. Jon Orens, world-renowned Johns Hopkins pulmonary medical transplant legend) and Firefighter Jimmy Dillon (now retired), FDNY lieutenant and former chief of the M-LFD who has followed in his the footsteps of his dad, the late FDNY Deputy Commissioner Thomas J. Dillon. (Photo by Russell Randolph)

John Alston

Are You Useful or Useless?

As a fireground commander, Battalion Chief John Alston definitely "gets" the incredible responsibility of deciding where to send his troops. I've had the privilege to get to know John at FDIC the last few years. John clearly has become very popular at the national level. His commonsense approach, humor, and down-to-earth attitude are infectious. He often focuses on the younger up-and-coming firefighters and officers—a great example of what so many of us need to do just as soon as we have stuff to give back. One of my favorite talks John has given was about usefulness and being useless—our responsibilities for knowing each tool and task we must perform when arriving on the scene. John often states that, from the start, it's important to desire, inquire, aspire, and retire. Hopefully John won't retire any time soon—so we can all gain.

John Alston

The Seven Cs

Dear Colleague,

I have been fortunate to work with a number of fire officers, from every rank and in many different departments. It's a privilege and an honor to train new officers in a variety of ranks and disciplines. I count it as a privilege because it gives us all a chance to share, learn, and gain insight into the many components of command. The prevailing question on their minds has always been; "How do you know when you know?"

These seven Cs of fire officer trust (commitment, competence, confidence, communication, courtesy, consistency, and courage) can follow in the order that I have given them; or may follow what best suits you and your situation; save the last . . . *courage*.

COMMITMENT

It's all about commitments. Notice I say commitments—plural—because there are more than one. It begins with a true commitment to yourself; the desire to be the best that you can be through preparation and practice. You strive to do all that you can to gain the right knowledge and acquire the pertinent skill set for your position. Then there's the commitment to your families, making sure that the trust they have placed in you (while you spend time away from them) was not in vain. No one forced you to select this career. You honor their faith in you by committing to be and do your best. Then there's the commitment to your agency or organization. Whatever type of organization it is—volunteer or career—you must commit to give them one thousand percent of your efforts, required time and resources. You must be one thousand percent committed to conforming to the rules, regulations, and policies of that agency or department. You must be committed to the personnel you supervise, your cohorts and colleagues, and your superiors. You must also be committed to caring for the equipment that has been placed in your charge. And you must be committed to the vision, mission, and goals of the organization. Then you must make a conscious effort to be committed to the citizens that your organization serves, to be professional at all times, and to provide the best possible service you and your organization can deliver.

COMPETENCE

Competence means continuing to learn and grow in your organization. Learn the inner and outer workings of it. Learn and work with other agencies that may interact or support your organization. Learn and master your policies and procedures. Take courses that increase your knowledge base. Seek out opportunities and events that allow you to share and network with others in your profession, others who may know a better way of doing things and sometimes do not share your same ideas or views. Stay current. Read the trade publications. Search the Internet; attend trade shows and training seminars. Expand your awareness of industry trends and evolving technologies that can assist you in the effective management of your responsibilities. Master the tactics and tools, strategies and rules of your organization. Know them cold . . . and *read, read, read, read . . . read!*

CONFIDENCE

With your commitments identified and your competence improving through reading, studying, and expanding your knowledge base, you can focus on the third C, which comes automatically: *confidence*. Confidence comes when you know who and where you are in the table of organization. It comes when you know your function and purpose. Confidence comes from knowing your responsibilities, knowing to and for whom you are responsible, knowing the length and breath of your authority or purview, and knowing the depth of your commitments.

When confidence comes, it gives you *command presence*. This is not to be confused with ego or evolve into arrogance. Confidence does not have to be boastful or egotistical, but rather can and should be quiet assurance of what is *right, fair,* and *appropriate*.

COMMUNICATION

The fourth C is one of the most pivotal: *communication*. Effective and appropriate communication is critical at all times and at all levels in our profession.

In our offices and fire stations, written, visual, and other nonverbal communication affect day-to-day operations. On the fireground or the scene of an emergency, communication by radio, mobile phone, material safety data sheets, preplanned guidelines, and computer data terminal is imperative. It is important, when managing people and emergencies, to effectively convey thoughts, orders, and concerns. Communication is the bedrock of how we get things done, yet many times its significance is overlooked. Making sure that we communicate effectively is *job one*. Communication can always be improved, and the study and practical application of conveying messages must be learned and practiced. The only necessary component we do not have control over is *feedback*. Feedback comes from listening. Communication is considered by many to be a two-way process, but I think often it is a three-way process because when a message is sent, there is the *sender*, the *message*, and the *receiver*. We must be clear with our messages; we must be effective with our communication style; more importantly, we must listen for confirmation and/or questions regarding the message. We have to be mindful of our delivery system. We have to be aware of our surroundings, interference, noise, static, and/or perception. We must study communication and practice communication. We must seek opportunities to determine if our message was transmitted correctly. We must also be patient when we are listening. Hearing is the physical act of receiving the sound. Listening is interpretation and processing. There is a difference (ask any married person).

To attain fire officer trust, we must be just as good at listening as we are at communicating. *Seek to hear, before you are heard!*

COURTESY

On September 13, 1981, my dear mother left this earth. Not a day goes by that I don't miss her. But the lessons I learned from her and my dad still hold substance. Through it all, I remember what they taught me, and I honor them by exercising it: *courtesy*.

She had simple rules:

- Keep your hands to yourself.
- Share and share alike.
- If you don't have anything nice to say about someone, don't say anything at all.
- If you have the ability to help someone, anyone, you help.

That's what we do in our business. We help people. They call us, we show up, and we fix things. If we can't, we get someone there who can, and we don't leave them until they do. Be *courteous* at all times! It makes a difference. Our profession, in some places, is getting a bad name, undeservedly so and in some cases we earned the bad "rep" from the actions and attitudes of some of our own. *Be professional.* It costs you nothing and pays immeasurable dividends.

CONSISTENCY

I have heard over and over again that the major difficulty with fire officers and leaders is consistency. The person in question acted one way as a firefighter and then changed upon becoming an officer. Rarely was the change positive. For example, as a firefighter, the person never wore the uniform or PPE correctly, but now as an officer requires subordinates to, with the threat of discipline for noncompliance. Be honest, if you sought the position for the rank and pay, say so. Most good officers were consistently good firefighters to begin with. The fire service and firefighters in particular have long memories. If you were shaky or bull-headed as a firefighter and think that everything is on the level now, and that members have to respect you because of your rank, good luck! Your days are going to be long and unpleasant, to say the least. At the worst, you are not providing the best service and are increasing the potential for someone getting hurt. Be consistent.

COURAGE

The final C stands for *courage*. Change the things you can. Accept the things you can't, but have the *courage* to try! Courage is not just needed on the field of engagement. It's needed in the decision making of our organizations. We have to have the courage to change our culture, ourselves, and our thinking when it is needed. I have witnessed some heroic and courageous lifesaving acts and decisions on the fireground that still impress me to this day. Yet, I also witness day in and day out officers who lack the courage to insist that their members wear seat belts and operate equipment safely.

Courage is needed, sometimes, when we have to admit we were wrong or that we failed to meet our objective. The *cowboy way* is not always the courageous way.

IN CONCLUSION

You don't have to agree. I know these things work. Try all of them. I hope. Try one. It can't hurt.

The answer to the original question, "How do you know when you know?" is: You will know when your peers and community respond to you in such a way that you know they trust you. It will be palpable. Until then, keep striving!

See you out there!

Joseph J. Apuzzio

on behalf of the family of FF/EMT Kevin A. Apuzzio

The Father of a Fallen Hero Firefighter

I met Dr. Joe Apuzzio the week his son Kevin was killed in the line of duty. While I try to assist and engage in all LODDs, this one was special due to the longtime friendship I had with the East Franklin, NJ, FD and its chief. On April 11, 2006, Dr. Joe's son, FF/foreman Kevin Apuzzio, died while he and four members of the East Franklin, NJ, FD selflessly and heroically attempted to rescue 75-year-old Betty Scott from a single-family dwelling fire. Kevin died when he became trapped after falling through a floor that collapsed into a well-involved basement, just a few feet from the front door—with Mrs. Scott in hand. It is sadly but widely acknowledged that most firefighter LODDs can clearly be prevented. However, in this case, it is well documented that Kevin died while he and his crew were heroically attempting a rescue—that based upon conditions and size up, it was the right thing to do. While most of us have had the opportunity to learn about those who have given the ultimate sacrifice from reading the reports, it's an honor to have my friend, Dr. Joe Apuzzio, share his thoughts on the loss of his son.

Kevin's father, Dr. Joe, on left, Marili, Kevin's mom, Leila, Kevin's sister, and Kevin

The Importance of the Support of the Brotherhood of Firefighters to the Family after a Line-of-Duty Death

It has taken me more than six years to write this article since the line-of-duty death of our son, FF/EMT Kevin A. Apuzzio. Kevin was a volunteer firefighter who died while trying to save a woman from her burning home. For several years it was very difficult for me either to speak or to write about anything concerning that dreadful day. Not because I didn't want to; I just could not bring myself to do it. If I had the opportunity to speak, I became choked up and unable to talk. So now my purpose in writing this article is to thank the firefighter service, the members of the East Franklin Volunteer Fire Department Station 27, Somerset, New Jersey, the Union New Jersey Fire Department, The National Fallen Firefighter Foundation, and many other firefighters and fire services for their continuing support to our family. I also want to make others aware of the importance and significance of the brotherhood's support for the families of firefighters lost in a line-of-duty death (LODD).

From the day that it happened, Tuesday morning, April 11, 2006, our family has been blessed with the support and friendship of the members of the fire service, of the East Franklin Volunteer Fire Department Station 27, Somerset, New Jersey, and the Union New Jersey Fire Department. From that first day there was always a fire officer with us and stationed outside our home from morning until dusk. It was reassuring for me to look out my bedroom window early Wednesday morning—the day after Kevin died—and see Chief Jimmy Davitt of the Union New Jersey fire department there from early in the morning until late at night. Chief Davitt quickly became part of the Apuzzio extended family. He took care of many things for us, tasks that we couldn't even think about doing in those early days following Kevin's death.

In dealing with the grief of our tragic loss, we and families with a similar loss have difficulty proceeding with our usual daily lives. Chief Davitt was instrumental in many ways in helping us along the path toward recovery. For example, he was our family spokesperson in handling the news media at that time and also at the time of the funeral and at Kevin's graduation from Rutgers University.

The heroic death of Kevin was a major news media event for all television and radio stations in the New York Metropolitan area. Kevin was a 21-year-old fourth-year college student studying criminal justice at Rutgers University in New Brunswick, New Jersey. He also was a certified EMT and volunteer firefighter. What a news story. There was an editorial written two days after he died in the Star Ledger, which is the newspaper that has circulation throughout the state of New Jersey. The editorial described him as a hero and a role model for young people.

There were numerous other inquiries from the news media to interview our family members. However, one of the last things that our family wanted to do at that time was to speak to the news media or be on television news. We wanted privacy. Chief Davitt was our spokesperson from the time of death through the funeral. In addition, in late May 2006, Kevin was granted his degree and graduated posthumously from Rutgers University, New Brunswick, New Jersey. Again there were many members of the news media present, requesting an interview with us, and again Chief Davitt was our spokesperson. He did an excellent job so our family could have privacy during our grieving process.

Kevin's funeral was a fireman's funeral. It was so large and attended by so many that the wake could not take place in a funeral home. So with the agreement of Pastor Father Charles McDermott, the wake and funeral was held in St. Michael's Church in Union, New Jersey. This was the family's church for us all, so it brought some comfort to us.

It was estimated that there were more than 5,000 attendees, including the governor of New Jersey and the president of Rutgers University. Most attendees were firefighters from around the country who came both to the wake and funeral to honor their brother Kevin. It was a tremendous show of brotherhood support, not only for Kevin but for us as family. I cannot, even to this day, explain the importance to us of seeing all the firefighters in attendance and their condolences to us.

Since then, the East Franklin Volunteer Fire Department Station 27 in Somerset, New Jersey, has become our second home, as it was Kevin's while he was a Rutgers University. Our family has been adopted into the brotherhood and family of firefighters. I even became an honorary member of EFFD. My badge number is 2713, the same number that Kevin had.

Chief Dan Krushinski and his wife Lisa have made us a part of their family as well. Since Kevin's death we've attended events at East Franklin including Christmas parties, the annual installation of officers' dinner, and even the Sweet 16 birthday party for Chief Dan and Lisa's lovely twins. Chief Krushinski allows me to administer the oath of office to the EFFD firefighters at the installation of officers each January. It is an honor for me and to the memory of Kevin to do so. Our family looks forward to going to the firehouse for these events and others such as the Memorial Day celebration, and Christmas and New Year's parties.

Just outside the EFFD firehouse stands a life-size bronze statue of Kevin. The statue was built and erected with the funds from many donations from the public and the donated time of the construction crews, many of whom are firefighters. My wife and I often travel to the firehouse on weekends just to spend some time at the firehouse and look at the statue.

In closing, the Apuzzio family would like to thank all of firefighters for their continuing support, the Union New Jersey Fire Department under Chief Fred Fritz, the members of East Franklin Volunteer Fire Department Station 27, under Chief Krushinski for making us a part of their extended family, the Fallen Firefighter Foundation, and many others.

I have said several times at a gatherings honoring Kevin that after his death I needed and was very fortunate to have the brotherhood of firefighters support. Two of the most important people to our family served as "crutches" to guide me for a while until I "got back on my feet." One was Chief Jimmy Davitt who was at our house every day right up through the funeral, and the other was Chief Dan Krushinski. These men and their families have supported us and continue to support us along with all the members of East Franklin Fire Department to this very day.

I'm not sure what would've happened had we not had the kind of support we were given; however, I do know that without that support, our family would have had a much more difficult time dealing with the loss of Kevin. The support of the members in the fire service and EFFD has helped us immensely on the road to healing.

STEVE AUSTIN

On the Roadway—He Has Our Backs

As an old friend, I have enjoyed "watching" Steve expertly maneuver the national political scene. There are few elected officials whom Steve doesn't know how to reach out to, and he understands how politics work. For example, Steve is a fire collector (as well as a veteran firefighter and fire police officer), and in his large firehouse-like garage, he has hundreds of photos of himself with well-known elected officials, including numerous presidents. After looking at the photos for a while, you will notice one thing—the Democrats are in one section and the Republicans are in another. Ahh politics! In the last few years, Steve has quite possibly had his biggest impact in protecting us on the roadways of North America. Through his leadership working with the Cumberland Valley Volunteer Firemen's Association, they have launched "Responder Safety," the only website 100 percent focused on our ability to survive while operating on the roads. Steve has been a "giver" his entire career—and once again, Steve is giving back.

Steve Austin in his restored fire truck

What I Have Learned

For a fire department to function properly there needs to be a mix of the enthusiasm of youth and the wisdom of age. Like a good broth, the mix can't be too sweet or too sour, too spicy or too bland. Good leadership, like a good chef, makes all the difference.

The center of the fire service universe is not the day room of your station, and the rest of the world does not revolve around your department. The universe has an order; to understand the fire service universe, you must venture into lands unknown and uncharted. You'll be surprised to find there is intelligent life out there.

Individuals and individual fire companies cannot enact change as quickly or as effectively as an organization of likeminded people with diverse ideas and opinions can. Strong volunteer associations, local unions, and regional and national organizations deserve your support and participation.

Fire service tradition is the glue that holds everything else together. Tradition should not be scoffed. Neither should tradition be used as an excuse for rejecting new ideas.

A clean station and clean apparatus says more about your department than any PR campaign can.

No matter how long you have been in the fire service, you cannot have too much training. When we think we know it all, it's long past the time to retire.

Many of us don't know when to say when in the fire service. This can lead to personal harm and to the detriment to the public we serve.

Elected and appointed officials don't owe us anything. We have to prove our worth every day in order to gain their trust. When we earn that trust, we shouldn't abuse it. The fire service isn't the only constituency we were elected to represent. Don't take our friends for granted.

Poor firefighter behavior can wreck a department's public image in a heartbeat.

The differences between the career and volunteer fire service are not nearly as great as their similarities. The mission is the same. Attacking one another is debilitating and pointless.

Being a firefighter defines who you are. If you are not proud of being a firefighter, consider doing something else.

A successful firefighter has the right mix of formal education, vocational training, and common sense. All are required to be recognized, promoted, and ultimately respected by peers.

The family of a deceased brother or sister is grateful long after the funeral when you call or drop by to ask them how they are doing.

It takes guts to take people aside who are committing unsafe acts on the fireground and tell them you do not want to see them hurt.

Nothing is wrong with saying hello to everyone in the firehouse when you walk in. You don't have to like them enough to take them out to dinner, but the least you can do is speak. This is a brotherhood, you know.

The color of emergency lighting on apparatus is the third rail of fire service politics. Do yourself a favor and stay out of that controversy.

Being called a *firefighter* is the greatest compliment you will ever receive.

Anthony Avillo

"Get'm In Safe . . . Work'm Safe . . . and Get'm Out Safe"

Deputy Chief Anthony Avillo of North Hudson Regional (NJ) Fire & Rescue has authored several books on fireground operations and has been an FDIC instructor for many years. He is probably best known for being a fireground "strategist," teaching us the realities of the varied scenarios we encounter. His understanding of command, control, and discipline on the fireground along with his shared mantra of "get'm in safe, work'm safe, and get'm out safe" has set a tone nationally as the service continues to understand the risk/benefit factors of doing the job. If any fire officers are confused about their role and responsibility of keeping their personnel safe, Chief Avillo will help them refocus.

La famiglia—the reason I want to make it home after every tour

The Failure of Imagination and the Unintended Consequences

I chose this topic because most of the bad things that happen in this business are preventable, if accurately analyzed before they become an issue. What exactly are unintended consequences? They are the random results from actions or inactions that are detrimental to the mission of the fire service. The guilty party is not necessarily always an officer, but may be the department itself, including its leaders and administrators. In addition, the victims are not necessarily those who have initiated the unacceptable actions or inaction. When the randomly unthinkable happens and it could've been prevented by addressing a situation beforehand, it is often termed as a "failure of imagination." Nowhere within a department or in an officer's or individual's mind was this set of circumstances and consequences foreseen. Unfortunately, this is the case with most unintended consequences.

An officer confronted with a situation, whether in the hard environment (the emergency ground) or the soft (everywhere else, including the nothing-showing environment), must not be a victim of imagination failure. When we take the time to consider the possible consequences and always display a willingness to do the right thing all the time, random consequences can be prevented, thwarting the severity of imagination failure, if not preventing it completely. If you are not going to conduct yourself properly in the soft environment and practice consequence management, it will never happen on the fireground . . . and both the imagination failures and the consequences will be even worse.

Take, for example, a real-life situation that occurred during a multiple-alarm fire. A captain got caught in an aerial device and was severely injured. The situation was as follows: A heavy fire condition existed on the top floor and cockloft of a five-story multiple dwelling of ordinary construction. The incident had turned defensive, and the ladder pipe was being utilized. A firefighter had climbed to the top of the aerial to direct the stream. Someone on the ground was not comfortable with the stream or the operations, so a captain was told to deliver the message and he climbed up the aerial to address it. This was necessary because the intercom at the end of the aerial was out of service and had been for some time. The department had neglected to fix it so there was no turntable or ground-to-aerial tip communications available.

This fire took place at a time when not everyone carried portable radios. The captain got to the top and surveyed the situation. At this point, he motioned for the aerial operator to lower the aerial, possibly to get a better angle at the fire. Instead, the aerial operator retracted the aerial and caused the rungs to clamp onto the foot and ankle of the captain as they began to retract. This caused severe injury to the captain. What was worse was that the aerial device hydraulics locked up and now the captain was stuck in the aerial five floors up. Believe me, the way he was yelling, we didn't need an intercom to hear him. If you remember how an older aerial was operated when the hydraulics failed, it was by a manual hand crank. You had to physically crank the aerial back to retract it. The problem was, which way do you turn the crank? Well, if he screams louder, you know you are going in the wrong direction—true story! Anyway, the damage to the captain's foot and ankle kept him off the job for over a year.

So why was the intercom not fixed? Was it because it was "just an intercom?" What was the hierarchy of importance in fixing that intercom? Could the department have foreseen consequences such as this from not fixing the intercom in a timely fashion? Of course not, no one could . . . another case of failure to imagine. On this day, the gamble of not fixing it ran out. Thus, the moral of the story is this: Don't wait until the unforeseeable suddenly comes into view. Chances are you will not like the consequences, and the loss that occurs is random. In this case, failure to fix the $100 intercom cost thousands in overtime, medical bills, and apparatus repair, not to mention (and most importantly) the loss of service of a veteran company officer.

The consequences of imagination failure are not only the fault of the administration, but are many times the fault of the line officers. Consider this scenario: An engine company responds to a motor vehicle accident (MVA) in their first-due district in the early morning hours on a Saturday. They have only an hour left in the shift. They need to utilize speedy dry to soak up a large fluid spill and use all that they have on the apparatus. As the shift is ending, the off-going captain passes on to the oncoming captain that they are out of speedy dry and he will need to replenish the inventory that day. The department keeps the reserve inventory in one of the firehouses of a different battalion, and the duty of the company is to secure speedy dry during that 24-hour tour of duty. The captain failed to get speedy dry that day because it is in another battalion and, after all, it's just speedy dry. The night passed and the next morning as the shifts are changing, another alarm comes in for the same area for an MVA. Again, there is a need for a significant amount of speedy dry but, because the captain did not get it the day before, the oncoming shift is left without it. They have no choice but to request another company to respond with speedy dry. This brings an engine company from another district to the scene. While they are there, an alarm for a reported fire with possible people trapped comes in for the first-due area where the second engine just vacated to bring the speedy dry down. Now there's a possible delay in water and no support for a rescue operation should it be required, all because the second engine is out of the area to do the job that the first engine should have handled. Do you think the officer who did not get speedy dry foresaw the possible unintended consequences of his inaction? Was this laziness? Was this a failure to see the bigger picture? To be honest, it is all of those reasons, and it doesn't matter to the person trapped in the fire building. All it took was a little effort and care to do the job properly rather than pass it on to someone else, in this case, the relieving captain.

At all times, you must do your job, keeping in mind the unintended consequences of your actions. This includes addressing sub-par performance on your watch. As such, there are three phrases that should never be in the officer's vocabulary when addressing an issue that requires attention. These are:

- "It's no big deal."
- "It's okay."
- "Don't worry about it."

How many officers do you think have regretted saying these words to a subordinate? These statements often get uttered at the same time (usually as the end statement) in a fix-your-people statement. It might go something like this, (superior to subordinate) "Regarding the way your apparatus was positioned at the last call, you should have been up a little further so the ladder had more room, but it's no big deal." Another example is, again superior to subordinate, "Your companies were not in proper PPE on the last alarm. That's not department policy, they should be geared up . . . but don't worry about it. It's okay." Do these statements sound familiar? Don't send out mixed messages to your people. If there is something you are not comfortable with, let them know. If you don't, they will think you don't care. Failure to care is the absolute worst trait an officer can have in this business. It destroys both credibility and trust. In addition, let them know why it is not okay and explain your expectations regarding future actions. Make sure you have agreement and an understanding so that both parties understand the same thing: your expectation.

A consistent officer (the absolute best trait) always addresses the uncomfortable and always looks at the potential for unintended consequences, allowing his or her imagination to see both the possibilities for tragedy and the opportunity for success.

Stay safe out there.

Douglas Barry

I've gotten to know Doug over the last few years, serving with him on the board of directors of the National Fallen Firefighters Foundation. Doug recently retired as the chief of the Los Angeles Fire Department—an intriguing position given the numerous challenges. Doug spent 34 years on the job as a firefighter, apparatus operator, engineer, captain, battalion chief, chief of staff, assistant chief, assistant fire marshal, and retiring as the chief. He worked in and commanded some of LAFD's busiest stations in South L.A., Port of L.A., LAX, and the Wilshire Corridor. As a chief he commanded fires but also oversaw management of department discipline, wellness and risk management programs, served as liaison to the city attorney's office, LAPD, city council and mayor's office, and much more. Doug is another one of the contributors to this book who values higher education as well as community service outside of the fire service. His life of service provides him with much to pass on to us.

Douglas Barry

Fire Service Leadership

Leadership has been much discussed and written about in the fire service. In the end we all know what leadership looks like, but often find it difficult to describe exactly what it is. To me, leadership is best described as instilling a sense of confidence in others to the extent that they willingly want to follow. This does not necessarily mean they always like the direction the leader want them to follow, but in spite of their reservations, they have confidence that the direction is ultimately the correct one.

In my view effective leadership is undervalued in our society and is often taken for granted. It seems great leadership is not recognized until it is demonstrated during a crisis such as a war or other tragic events. But effective leadership is important can and must be demonstrated daily, especially in the fire service. Shrinking budgets, devastating natural and man-made disasters, and balancing traditional values with progressive thinking challenge today's fire service leaders.

Some say that leaders are born; others say they are made. I say both are true. Innate personal qualities such as command presence and extreme self-confidence come natural to many great leaders; however skills such as strategy development and critical thinking techniques can be learned and developed.

The purpose of this article is to discuss three aspects of leadership that are often overlooked and to provide insight to new and prospective leaders on how they should be viewed. The concepts are leadership courage, integrity and credibility, and the loneliness associated with being a leader. All leaders experience these to some degree; therefore, it is important to recognize their significance.

LEADERSHIP COURAGE

Fire service leaders are often required to make life-and-death decisions. These decisions require a tremendous amount of courage and self-confidence because they are often second-guessed by others. Whether it is making the call to evacuate firefighters from a building or to make the unenviable task to close fire companies, the courage necessary rivals that of fighting the most dangerous fires.

Great leaders come to realize second-guessing will always exist. They make their decisions based upon their training, experience, and intuition with little regard to how the decisions may be viewed after the fact. This takes great self-confidence and a thick skin. Though most great leaders realize they are not infallible and have made decisions that they later would like to change, they embrace their infallibility and continue to lead with confidence.

LONELINESS OF LEADERSHIP

Consistent with leadership courage is the concept of the loneliness of leadership. As a leader you have the ultimate responsibility for all that occurs under your command. The buck stops with you. Those without that responsibility can and do give their opinions on what they would do if they were in charge, but only the one shouldering the responsibility can truly know and feel the burden that comes with it.

Self-confidence and the acceptance of the realization that no one is infallible is helpful in combating the feeling that you are alone. Also, confiding in colleagues with equal responsibility gives a sense of kinship. Just knowing that other leaders are facing the same feelings and emotions can be helpful in dealing with this common sentiment.

INTEGRITY/CREDIBILITY

It is important for leaders to realize that a major part of their ability to influence others is closely linked to their integrity and credibility they exhibit. Demonstrating integrity in all aspects of leadership breeds credibility which can pay huge dividends when negotiating with employee organizations, with elected officials, and with the public. It can open doors that otherwise would be shut and minimizes potential challenges and doubts that would normally exist. Integrity should enter into every aspect of a leader's decision-making and should not be compromised or taken lightly. Integrity and credibility are a leaders most valuable commodities and when lost they are nearly impossible to get back.

CONCLUSION

In closing, there is one last thought to leave with you. Every leader leaves a legacy. Whether you do a lot or nothing at all, your legacy is remembered long after you leave or retire. The important thing to consider is that we all want to have a long-lasting positive impact on our commands. We want to make things better for the present and future. Because of this, all leaders should be conscience of the possible legacy being left as a result of their leadership. They should strive to commit themselves wholeheartedly to the responsibility of their commands and leave a legacy of a strong productive effective leader.

MARC BASHOOR

Leading, Supporting, and Responding at Multiple Levels—With Proven Success

Most people who might take time to even attempt to understand the job of chief of the nation's largest and busiest combination fire/EMS department would be perplexed at the wildly diverse roles, runs, and responsibilities . . . as well as the politics. Challenging? Ya think? Time consuming? Uh, yeah. A 9–5 job? *Fahgetaboutit.* Rewarding. *Far beyond that.*

Leading the Prince George's County (MD) Fire-EMS Department's 2,000 career and volunteer members (in 40 stations) requires all the standard words such as dedicated and committed. But moreover, *this* job requires defined enthusiasm and a love for the job, the community, and especially the people doing it. Chief Marc Bashoor loves it 24/7/365. Be it running laps with recruits, handing out smoke detectors, visiting with the families of injured firefighters, riding the rigs with personnel, being involved in the neighborhoods, or responding to support operations at working incidents, he is literally all over it. And he passes it on by communicating "what's up" through (for example) Twitter; he communicates several times a day to his members, the elected officials, and the community about what's going on.

Starting as a volunteer firefighter and working his way up within the department, Marc Bashoor seems to have struck a great balance in clearly allowing his career and volunteer officers to run the day-to-day operations—but when the chief needs to be seen or to lead, he is there, be it operationally, politically, or in the community, representing "their" fire-EMS department

Leading Forward—A Fire Chief's Expectation

Let's face it—almost anyone can put out a fire or stick an IV. They need to be well trained and have the right tools and a willing fire or patient—and I'll admit, in some cases a little luck doesn't hurt! Likewise, well trained commanders can lead most scenes to successful outcomes—with enough crews that can put a fire out or stick an IV successfully. In my experiences, I've seen the "average" firefighter or paramedic on the street use the "good fire" or "good stick" measures to determine the outright effectiveness of their department. We know this is shortsighted—even though "we" were probably one of "them" at one point in time.

EXCEPTIONAL LEADERS

Clearly the breadth of any department requires a much broader and more dynamic examination of detail to determine the department's level of effectiveness—with the ultimate measure of ensuring that we make a positive difference for our communities. One of the most significant challenges "exceptional" leaders face in making a difference is the management of *change*. It may be safety- or standards-based, contractual, volunteer-career issues, financial—whatever it is, "exceptional" leaders needing to make the change need to "lead forward." I absolutely believe in the practice of allowing lower levels of leadership to manage their areas and make their internal differences one "move" at a time. However, the "exceptional" leaders I'm talking about will "lead forward," personally "selling" the programs or engaging the community, politicians, or leaders of other organizations in active participation to achieve the necessary change. In some cases the change is something the majority wants, but can't seem to get to. "Exceptional" leaders are able to build consensus and navigate the labyrinth of policy and punctuation to bring their organization along—"leading forward," then having other department managers "carry the torch."

TRUE LEADERS

I said before it is "easy" to put out a fire or stick an IV, and that "exceptional" leaders can effectively manage by using good people and manage change by leading forward. It is important to note that it is *not* easy to be a "true" leader. The difference between an "exceptional" leader and a "true" leader is measured by that leader's ability to take an organization where it *ought* to be, as opposed to where it *wants* to be. You can call it "leading by example" or whatever you'd like—in the end, the "true" leader will be out in front of change, whether personally or in program initiation—whether popular or unpopular, easy or difficult, large or small. Consensus may not be easy to come by in true leadership positions—in fact, depending on the number of facets involved, consensus may be unattainable. This is where the "true" leader is really tested. Keeping your "train on the track" while there are multiple attempts at derailment will build character and test the resolve of many "true" leaders.

The National Fallen Firefighters Foundation's (NFFF) 16 life safety initiatives include the "Courage to Be Safe" training program. The National Institute of Standards and Technology (NIST) and Underwriters Laboratory (UL) studies of fire flow and flow paths is groundbreaking scientific research. I submit to you that the NFFF initiatives and the NIST/UL research are examples of "true" leadership. They are initiatives that might in many cases not be popular, are not "sexy," and at times bristle up against the bravado we know as "tradition" in the fire service. Not only does it take courage to be safe, it also takes courage to be a "true" leader.

The American fire service is steeped in that tradition and pride—although I am the first generation fire service member for my family, I am a proud 32-year product of that tradition and pride. We *must not* allow the old adage "100 years of tradition, unimpeded by progress" be the reality. Much like horse-drawn wagons, tin helmets, leather coats, rubber boots, back steps, and the myriad other improvements, this much is true—change is coming every day. If we are going to make a *true* difference for the future—specifically in the rates of firefighter injuries and fatalities—our leaders need to be "true" forward leaders. We will save more firefighters and civilians and save more property if our "exceptional" leaders become "true" leaders—by taking us where we *ought* to be and need to be, instead of where we want to be.

"Do as I do." Chief Marc Bashoor joins recruits for their physical training run.

Phil Bird

When They Want to Be Like You

We often hear people use terms related to wanting to be like someone. In most cases, that's quite an honor—to hear that someone was so impressed with the way you act, behave, and lead, that they want to *be* that person. A few young firefighters whom I am very close to told me about Phil Bird. These young kids went through career recruit school a few years ago upon being hired in Prince George's County (Maryland) Fire-EMS Department. This group of probies was fortunate to have a solid group of no-nonsense instructors—and one of them was Lt. Phil Bird, an instructor detailed to the academy from one of the busiest truck companies in the nation. His attitude, enthusiasm, and professionalism stuck with one particular recruit—and all the recruits in that class. As fire chiefs, we sincerely hope our company officers are viewed in that same manner. In my opinion, it all starts and ends with company officers. They make it or break it. Company officers are where the rubber meets the road—and where every fire chief's goals are carried out. You can have a bad chief and work can still get done, but if you don't have solid company officers, chiefs can kick and scream all they want, the stuff rarely gets done. The link to success in any fire department are the company officers. Here are some thoughts being passed on to you from a pretty good one.

Today's Fire Officer

My name is Phil Bird. I currently serve as a fire lieutenant for Prince George's County, Maryland. I got my start as a volunteer firefighter in 1991 in Lancaster, Pennsylvania. I'm a first generation firefighter who has always wanted to do this as a career from about the time I could crawl.

I moved down to Prince George's County, Maryland, in 1996 shortly after high school, in hopes that I could land the job of my dreams as a full-time firefighter. As I applied to several jurisdictions in the Washington, D.C., metropolitan area, Prince George's County was the first to offer me a shot. I remember specifically in my job interview only asking the panel for just a chance at the best job on the world, and they obliged.

I've been assigned to several busy fire stations throughout Prince George's County that border southeast Washington, D.C., where the high call volume and fast pace helped me prepare in becoming a better firefighter. I was fortunate being inspired by working with some exceptional fire officers. I used a lot of what they taught me to promote to the rank of lieutenant. After working in the Training Division and various other stations, I'm currently assigned in a role that helps newer officers with further development. You may be asking yourself, why is there a program in his department that has officer development? The answer is simply that I work in a department where attrition has occurred faster than we could hire new recruits. When experience starts to dwindle, you have to play catch-up.

So in observing, critiquing, and mentoring these highly impressionable people, I have some advice for the "new fire officer." First of all, you haven't arrived yet. Well-respected fire officers are always learning, even in their off time. They subscribe to various periodicals to enhance their knowledge, and they apply it. They view various tactics found on the Internet, then host roundtable discussions at the station with their crews. Some people define this as "Monday morning quarterbacking." I call it "game planning." Like a pro football team that watches tape on their opponent to gain the upper hand. Find "teachable moments" on every alarm you and your crew take in. Hey Rook, where were the basement stairs on the last run? They find a way to keep their crews thinking "game time" on even the most annoying runs we get dispatched on.

Phil Bird of PGFD Truck 29 takes a break after a box alarm for a working dwelling fire.

I once was observing a new lieutenant discuss a pre-plan for a commercial occupancy in his area. The lieutenant's first question he asked his firefighters was, "What's the most important thing we need to discover while conducting a pre-plan?" I was thinking, this is great, this guys gets it. Way to engage the crew! The lieutenant said, "Getting the phone number of a 24-hour contact in case we have to force entry in the middle of the night." I almost fell on the floor!

Sometimes there's a little apprehension on behalf of the "new fire officer" when it comes to, well, being an officer. They don't want to disassociate themselves from a group that they've spent the first several years of their career following. Here's the good news: you don't have to! You do however need to know how to draw the line. If you are completely upfront as to where you want the bar to be set for your company, hopefully they will buy in. This may take more than just you writing bullet points on a whiteboard; you may have to show them with your actions.

My point is this: you as the "new fire officer" directly impact not only your crew, but their families as well. Every day you should seek the "teachable moment" to better prepare your team for "game day." We owe it to our fire service mentors in continuing to serve with a true passion for the best job in the world!

REBECCA BOUTIN

A Company Officer Who Risked It All to Save One of Her Own

As with others in this book, I got to know Captain Becky Boutin through writing about an incident she was intimately involved in. Westfield, Massachusetts, Firefighter Steve Makos is alive today because of Becky's heroism. "It was just pure reaction, knowing that my friend my coworker was in that condition," Becky stated. "I knew he was in big trouble and I knew I just had to get him out of there." According to FF Makos, "In my heart, you know I can honestly say if it wasn't for Becky and my fellow firefighters that day, there's no doubt in my mind that my mother would have been going to my funeral. I couldn't have come much closer than I did to not making it out of the house alive." We are certainly thankful that Becky was able to do what she did and that she wants to share her words with you.

Rebecca Boutin

Never Stop Training

The call doesn't start with the dispatch. It starts with the 5 a.m. trips to the gym; with the countless hours spent driving across the state for specialized training; with understanding what exactly is expected of a professional firefighter and being willing to live up to it.

You don't go into the fire knowing what's going to happen. You just make sure you're ready for any possibilities. Recruit training is about acquiring the building blocks, specialized training assembles them, and continuing education allows you to build on that foundation.

You can debate whether fire department culture evolves from the top down or the bottom up. I say it's much simpler. It starts inside each firefighter, and it's a very personal commitment. *Strong firefighter* means more than lifting heavy objects. It also means a strong heart that can work long hours, standing strong when anyone above or below you ridicules you for your dedication and commitment to be smarter, safer, or better trained.

Instinct tells you to go back in and get the downed firefighter, but it's training that tells you how to do it safely and successfully. So without the training it's just blind instinct—and that's what gets you killed.

Never stop training. We owe it to our brothers and sisters because that's how we all get through our shift as safe as possible. And never forget our loved ones who watch us walk out the door to those shifts. We owe it to them to be as prepared as we can to make it through and come home whole.

There's no reason training should be anything but enjoyable. Firefighters are competitive and challenge each other on the shift level to get stronger; then challenge other shifts and make each other stronger. Next thing you know, the whole department is working at a higher level.

There was nothing special about what happened in the rescue that I was involved in. Everything I used to initiate and effect the search and rescue came from my basic recruit training. Now obviously you don't just complete your recruit training and then go back to the firehouse and sit on it. You continue to train and reinforce that knowledge. You train with your shift and department so you also learn each other's tendencies. It's always preferred to know how you're going to mesh together before you're in hazardous conditions and lose line of sight of each other.

Changing unsafe behaviors should be the responsibility of each member of the department. This should start with policy changes made at the highest levels of the organization, supported by upper management, adopted and enforced by middle management, and followed by line personnel. Safety should be everybody's responsibility at all levels, and each individual should be held accountable for his or her own safety. This will allow for a much safer environment at fire scenes.

Heart attacks remain to be the leading preventable causes of on-duty deaths to firefighters. Stress, exertion, and other medical-related issues, which usually result in heart attacks or other sudden cardiac events, continue to account for the largest number of fatalities. Regular physical fitness has been linked to preventing line-of-duty deaths due to overexertion or stress, yet 70% of fire departments lack programs to promote fitness and health. Most fire departments do not require firefighters to exercise regularly, undergo periodic medical examinations, or have mandatory return to work evaluations after a major illness. We need to take our cardiovascular health seriously, even if the fire service does not. This cultural change needs to start from within.

I love the new culture I see becoming routine in the fire service. New firefighters are coming to us eager to learn and train. They're eating healthier and maintaining their strength and cardio workouts. This new attitude helps me as an officer keep a shift together. Now our whole daily routine can incorporate the values of this new culture. We cook and eat healthy meals together. We plan shift workouts. Even driving around town is a chance to check out buildings. Preplan. Keep up with what's new—even more importantly, what's getting old and unsafe.

A simple training tool is to quiz each other while doing routine firehouse chores. Hydraulics is a great example of knowledge every firefighter should know cold, like you can recite the Red Sox starting nine. But how many of us really do? So find out the next time you're washing vehicles. Nothing motivates a firefighter like a little friendly competition.

If I could change something about the fire service, it would be that our industry standards become mandatory and not mere guidelines. We have one of the most dangerous jobs, yet nothing is mandatory. For decades, this has given us too long of a leash to get hung up on, and we have not done a good job of being our own watchdogs. How long have we known that we're understaffed in the majority of our nation's fire responses? But the National Fire Protection Association (NFPA) is an association and not an agency, so it lacks the bite to effect change. It's unbearably frustrating that we know exactly how the majority of our firefighters die every year, we know exactly how to prevent it, but then we're not given the resources to make it stop. There is no question that firefighter fatalities and injuries would be reduced if every fire department were able to follow the guidelines set forth by existing NFPA standards.

The fire service has made great advances in personal protective gear, apparatus, equipment, and training; all the while the United States Fire Administration statistics show recent averages of more than 100 firefighter line-of-duty deaths per year and 10,000 serious line-of-duty injuries. Our industry is the most knowledgeable and safest it has ever been, but we are losing lives every year due to largely preventable circumstances.

We understand the science of fighting fires, but we need to understand the science of people fighting fire. The fire service as a whole is heaped in tradition, and change of any kind is often met with resistance. Start making positive changes within yourself, and help to change the fire safety culture as a whole from the inside out. Our culture is strong, and we must embrace its strengths. We should never be ashamed or ignorant of the "brotherhood's" strength; at the same time we should also be willing to acknowledge the weaknesses so that we can prevent line-of-duty deaths.

Why are we only learning from the loss of life? People have to get killed for us to pay attention, and even then the changes often turn out to be temporary. When everyone goes home safe, please understand that happens because of the training available to every firefighter. Let's reinforce the lesson that training isn't just about keeping time checked off in the training log. It's about saving lives, including our own, and we all know that that's exactly why we're here. When stories have the happiest of turnouts, I'd hate to think that they'd be overlooked because it has no bagpipes playing at the end.

Rebecca Boutin and Steve Makos

Rebecca Boutin

Garry Briese and Oren Bersagel-Briese

Selfless Leadership

I got to know Garry Briese when he became executive director of the International Association of Fire Chiefs. We developed a good friendship, and while we definitely locked horns on several occasions, we always ended up closer. While among the most personable, caring, and intense people I know, Garry is also one of the smartest and most forward thinking. His mind is decades ahead of where we are. I remember when he proposed high-visibility safety vests for us to use on roadways 20 years ago. Everyone laughed at him ... back then. He was also one of the earliest national leaders advocating for terrorism preparedness for the fire and emergency services (in the 1980s)—and he has done extensive research and lecturing in those areas. And now look at us. That is only two of hundreds of examples I could give you. Garry has been a firefighter, medic, and even a national Distinguished Eagle Scout. Currently he is the executive director of the Colorado State Fire Chiefs and is the executive director (accepting no pay) of the Firefighter Cancer Support Network and serves on several boards of directors, including the National Fallen Firefighters Foundation. Like many others in this book, I simply am not allowed enough space to tell you everything I want you to know about Garry Briese.

Garry Briese, left, and Lt. Oren Bersagel-Briese

The Badge, Family, and Success

The badge is the ultimate symbol of the commitment you made to your fire department, to your citizens, and to the profession. You are able to wear it only after you prove that you are willing to go as far as it takes to get the job done; and the badge is a part of you for the rest of your life.

Even as the badge is symbolically placed on your left chest, over your heart, always remember that it is your heart that supplies the badge, not the other way around. And your heart? That belongs to your family.

In this profession, you will find yourself away from your family for extended periods of time, away during significant family moments, and often away when your loved ones need you the most.

In spite of it all, never put the badge before your family. Don't wait for a significant emotional event to reset your priorities. Always know the importance and priority of family, and in the end, you'll never want to say, "I wish I had done it differently."

Sure, there will be moments when you have to decide between attending a class or spending a day with your family. There will be times when you have to be at the station when your family would rather have you at home. And there will be times when the tones or pager goes off, and you have to fulfill that promise you made to the citizens and department. This is the life of the firefighter.

However, there cannot come a time when you place your fire service aspirations in front of your family needs. Make sure that your family is taken care of, that they have what they need, and that they are not searching for "that person who used to be around." It may be hard to hear, but the fire will go out without you, the EMS call will be handled, and the department will still operate in your absence.

How do you put family before the badge? By telling the truth and saying, "Thanks for asking, but I have a family commitment." This is hard, especially when it is something you really would like to do or to become involved with. But, with experience and watching other colleagues struggle with the same decisions, it becomes easier to understand what you need and don't need to do.

Like it or not, we are often judged by the extra tasks we do at the fire department . . . those special projects, that new spec committee, or our willingness to work the extra shift.

Say no too many times and you risk getting a reputation for not being a team player or not being willing to carry your weight. Say yes too many times and you learn what it is like to be the figure who wasn't there when your family needed you.

All of us fear that those judgmental thoughts will become a subsurface influence during our next evaluation or promotion opportunity. But that fear is proving to be less factual because time after time, when firefighters are asked what leadership characteristics they want to see in their leaders, compassion and balance consistently rank near the top.

Put family as a priority in your personal and crew expectation list, and communicate it clearly. When you share your expectations with your crew, especially with new firefighters, you can see in their eyes that they are telling themselves there is no way they will let family interfere with the tasks at hand. And that's okay. You gave them permission to put family first, and when the time is right, they'll understand that the opportunity to be with family is accepted and encouraged. This is an incredibly important step in their personal growth that allows them to better focus on their role as firefighters.

As you place family in the highest priority of life, so do the people around you. Lead by example and model the behavior. Show them that they don't have to sacrifice home life in order to be an outstanding firefighter and leader. Show them how to properly balance the requirements of both the fire service and family to create a blend that results in a thoughtful, caring, professional leader.

Not only keep your family first, but place the families of those around you in the same priority. Unless it becomes cumbersome or intrusive, encourage your firefighters to take phone calls from their loved ones, at least to ensure that there isn't an emergency and to plan to make a return call. When you show that you care about those around you, they will care about you and what you know . . . and then you have truly arrived at the sacred starting point of mutual respect, empowered followership, and healthy leadership.

Share stories about your family, and let them share about theirs. Listen, remember, and return conversation. A few minutes at the start of your shift will not take away from the goals you have set for the crew; in fact, those minutes are an investment in the success of your crew and will only serve to get you there quicker. When you care about people, they will care about what you are trying to do as a leader.

And when they truly care about what you are trying to do, there is no limit to how far you can ask them to go or how far they will go for you. This is the hallmark of a memorable leader.

Alan Brunacini

Stepping Wayyy Outside of the Box

While Alan Brunacini is easily the most recognized, highly respected, and well-known fire chief in the world, he is also the most approachable—and that is good for all of us. While a million "Bruno" stories come to mind, I want to share one from 1984—a perfect example of stepping outside of the box. While attempting to rescue a worker who was overcome by fumes while cleaning out a 10,000-gallon toluene tank, Phoenix Firefighter Ricky A. Pearce was killed and 14 others were injured when the tank exploded. Bruno was very public about the facts of the incident, how it happened, and lessons learned. My focus here is actually not on the event, but how it was handled. Sharing the facts and speaking out publicly was highly unheard of "back in the day." I had been a chief for 2 years at the time of the explosion and was amazed at what Bruno had done. I ran into him at a conference and figured I would ask him why he was so open. His answer was quick and simple: "The lawyers are going to figure it all out anyway, so we might as well share what we know so this never happens again." Logic from Bruno . . . pretty much like most of the stuff he says.

The photos of this beautiful, beloved fire engine can be found in Bruno's wallet, right *in front* of other photos.

Equipment #1489

I received an assignment to produce a short article on any fire service topic. The subject that I chose is a little unusual and very personal. What I will write about is one of my most favorite things in the world—my fire truck. Perhaps describing such a long-term project with a happy ending can bring a little light relief to what will be very serious, critical, and relevant subjects that I am certain that my learned colleagues will skillfully present.

On June 2, 1958, I joined the Phoenix Fire Department and was assigned to Station One in downtown Phoenix. The station was an old, two-story "central" station that housed an engine, squad, aerial ladder, hose wagon, battalion chief, and an on-duty crew of 23 firefighters. I was assigned to Engine One, which was a six-person company that was first due to the downtown business district. The company apparatus was an open cab 1952 Mack, L-Model 1500 GPM pumper (equipment # 1489).

When I walked in to Station One and first laid eyes on 1489 I experienced a significant emotional event—who wouldn't? A 1950s-era L-Model Mack has a massive vertical chrome grille shell, a hood that is nearly six feet long accommodating an oversize commercial engine, adorned with huge Gothic-looking round front fenders, a large chrome pump panel with nine gauges, and a front bumper a foot high; the vehicle is built heavy enough to drive through the gates of Hell. The truck motor is a huge unmuffled gas engine that would rattle windows for about a block. From the moment I saw (and heard) the truck I absolutely had a blinding flash that someday I would own it. I worked on the truck as a firefighter, engineer, captain, and battalion chief, and I parked my vehicle right next to 1489 for six years.

The truck was in front-line service as Engine 1 from 1952 to 1972 and in reserve status from 1972 to 1977. Soon after it went out of service as a reserve unit, it was declared a surplus vehicle and was part of a regular city auction where obsolete equipment was sold. I bought the truck in 1978. I stored the truck for the next two years while I constructed a small building next to my residence to house the rig. In 1980 we took the truck apart and began a 30-year restoration project. My beautiful (and very patient) wife would occasionally mention that a completely disassembled full-sized fire truck produced a massive amount of "parts" . . . somewhat of an understatement.

For the duration of the project, she basically lived in a house with a fire truck junkyard in the back yard.

The superstar of the restoration project is my almost 50-year colleague Robert "Hoot" Gibson, who was for many years the deputy chief manager of the Phoenix Fire Department fleet. Hoot is a master mechanic, metalworker, and paint expert and the most skillful fire apparatus specialist anywhere. Hoot's son Wayne has followed in his dad's footsteps and is a world-class restoration expert. He assisted with the paint part of the project and applied stunning gold leaf decorations. He has been an invaluable partner. Hoot started and finished the project and basically worked on it every weekend in between. A huge benefit to the outcome of the project is the attention and precistance to detail; this is a reflection of both Hoot and Wayne being absolute perfectionists

From 1980 to 2011, mostly on weekends, we all labored with the complete restoration of all the mechanical parts of the truck including: the frame and undercarriage, all the running gear, the drive train, the 1500 GPM Hale pump and the original 480 Hall-Scott six cylinder, 1,000-cubic-inch engine. Every mechanical part was replaced, reconditioned, or re-machined, so the truck is now essentially brand new. Also, every part of the truck received a cosmetic restoration starting at bare metal (my job), including complete body restoration and show-quality paint. The entire standard inventory of firefighting hand tools, nozzles, and hydraulic adapters are on board; there is 2,000 feet of 2½-inch and 600 feet of 1½-inch cotton fire hose that has been loaded in place. The rig is equipped with a top-mounted large deck gun and two large outrigger guns on either side of the rear. All of the hydraulic appliances are polished brass.

The completed truck now resides securely in its own small fire station. I now am by default commander (actually custodian) of routine maintenance, cleaning, and conducting frequent tours. We recently entered 1489 in a large national car and truck show and won first pace in the commercial class. My son John (retired fire captain) is our expert equipment operator and occasionally takes the rig for a spin around the neighborhood. The truck always attracts a huge amount of attention when it is out in public.

I have enjoyed every stage of the 1489 project to put the truck in its current completed condition, and I now get to polish and maintain it so we can show it off. Someday I might look around for a home in a museum for the truck so that we can share its glory with future generations of apparatus admirers.

Timeless Tactical Truths

Early in my career, I developed the habit of writing notes on what I saw, comments I heard, and thoughts I had that related in some way to understand my job as a firefighter. Most of these notes in some way had to do with firefighting and fireground operations. I thought I would depart from my usual column and talk about some of the old truths and a few recent ones.

YOU CAN ONLY SAVE THE SAVABLE

We must always operate on the fireground with the understanding that we typically inherit a tactical situation that is underway when we arrive. The fire has created physical damage/injury/death that is basically irreversible—simply, we cannot unburn the structure and its contents or uninjure the occupants the fire beat up (or murdered) before our arrival. A major challenge for us is to estimate how long the fire has been present and the effect it has already created. Being able to accurately evaluate *what time it is* in tactical terms is the foundation for establishing an attack plan that realistically separates what is lost and what is savable.

DON'T GO ANYWHERE YOU CANNOT COME BACK FROM

We must always operate with the awareness that our survival depends on maintaining the ability to go into *and* come out of the hazard zone. The logical and least painful time for us to evaluate the round-trip entry-exit profile of the interior layout is at the very beginning of the attack process. This up-front exterior size-up requires us to slow down a bit and consciously decide (before we enter) if the overall situation is offensive or defensive. The size-up must be based on an unemotional understanding that it is easier (and a lot more fun) to go into a hazard zone than it is to come out (which can be a lot less fun). When we decide if conditions allow an interior attack, we must then decide on the best entry/exit door that gets us into the high-impact rescue and fire-control points. When we figure out where we enter, then we must decide what operational and command resources are required, how our travel path must be protected, and if a tactical reserve (rapid intervention crew, or RIC) is in place. The old fire inspector said, "If you pay to get in, you pay to get out." We ought to apply that to interior firefighting.

AN ENGINE WITHOUT A SUPPLY LINE IS JUST A 500-GALLON TANKER

It is the responsibility of every engine company to provide its own uninterrupted supply of water. Sometimes the officer, particularly the first-arriving pumper, trades taking the time to lay a supply line for pulling quick attack lines. Sailing by the last hydrant at 35 mph commits that company to fire control with only onboard water. If the fire does not go out, it also creates the need for a desperate prayer that a subsequent-arriving engine can lay the plug passer a supply line. The only sound worse than a pump that has just spun away from the last water in the tank is the grumpy voice of your battalion chief asking, "What part of my last line-laying lecture didn't you understand?"

DO NOT DO THE WRONG THING HARDER

Operational action typically achieves a pretty quick effect. Simply, it does not take long to know if what we are doing is working. Many times an operating position requires multiple activities, such as placing an initial attack line; adding reinforcement with a backup line; and engaging some truck company support, such as forcible entry, venting, and opening up. The incident commander (IC) must adequately invest in each basic tactical position by assigning all the resources needed to create a concentration of tactical force that can solve (overwhelm) the problem in that place/function. Most of the time, these resources get the job done. Occasionally, they don't. When this occurs, the IC must decide whether to expand that position or go on to Plan B. When the current plan is not working, many times the correct response is not to increase the investment in that position but to move on to a spot/function that will have a positive effect. This requires us to create a strong initial response in an operating position but not to fall in love with our attack plan. We must always maintain the tactical agility to move our attack to a place where we can cut off and control rather than chase and be behind.

IF YOU MUST BURN DOWN A BUILDING, DO IT WITH CLASS

As we operate on the fireground, we had better figure out that what the front end looks like pretty much produces a picture of the back end. Sometimes that back-end picture is defensive. This means in plain English that the building is going to burn down. When this occurs, we must capture, maintain, and not lose control of ourselves personally and our resources operationally, even though the fire in the involved fire area is beyond our offensive capability. Just because the fire is out of control does not mean that we should be. The way we stay in control is by doing the following:

- Always operate with early and strong strategic and tactical-level command.
- Effectively deploy adequate operational resources within SOPs with a continuous tactical reserve in place.
- Integrate safety and worker welfare automatically and critically into all operations.
- Use effective tactical operations to limit loss as closely as possible to the area of origin.
- Protect and care for customers within an added-value service delivery plan.
- Critique the incident to determine the lessons learned/reinforced and put those lessons into a well-timed organizational action plan.

Consistently doing all this creates high-class firefighting and displaces the opposite.

Alan Brunacini

John M. Buckman III

Volunteering with Demonstrated Passion . . . Local, State, and National

About a hundred years or so ago, a relatively young man stopped at a meeting of the IAFC Volunteer Section and asked how he could get involved. Ha—I thought to myself, someone to load work onto. And then I rethought—he's probably one of "those" who wants his name on something but will sit in the dugout, not really wanting to do work. So, we gave him a shot at bat anyway . . . and since that day, John has consistently hit home runs. Constant. Reliable. Midwestern gung-ho type.

I've been friends with John since that day in the 80s, and we've remained close ever since. If I were to pick one word to describe John, it would be passionate. Like many of those in this book and most of you reading it, he *loves* the fire service—in particular, the volunteer service.

He was chief of the German Township Volunteer Fire Department for 37 years and is now the chief of the Indiana Firefighter Training System. John's accomplished a whole lot—and, for example, in 1996 he was selected as the Volunteer Fire Chief of the Year. He is also past president of the IAFC. Be it his efforts in creating that extremely professional volunteer FD in German Township, or the positive influence he's had in creating the first statewide Indiana training system, or his efforts nationally, John will be able to look back and know that if he ever leaves the service, it is certainly better than when he first found it. He has always been one to "pass it on." Not bad for a Midwesterner!

Creating a Survival Mentality

A LEADER IS ONE WHO INFLUENCES CHANGE

Firefighters have a long history of putting their lives at risk while serving communities. The future firefighter must take a proactive approach to survival.

There are so many risks to our health and welfare. Some of the risks, such as building collapse, have immediate consequences, and others, such as cancer, are the result of long-term actions. Risk must be analyzed prior to making strategic decisions that commit our firefighters to dangerous to life and health environments. In the future, firefighters will be asked to do more and more but with less resources. You will be requested or ordered to perform superhuman feats without consideration to your health and welfare.

Remember in the fire service, it's not a question of *if* but *when* will you be in a situation that will require you to be the best at your job. There is no instant replay or do-over in the fire service. It must be done right the first time.

I believe this mentality also comes into play in our everyday lives. In many cases, having a survival mentality outlook can mean the difference between success and failure, and in a critical situation it could even mean life or death! We all face severities in our everyday life, and the ability to overcome these difficulties will determine the outcome. So what exactly is a "survival mentality"?

My definition of "survival mentality" is, simply put, "to be prepared and have the confidence in yourself that you have the ability to overcome whatever adversity you might face." You will achieve that confidence when you know in your heart and mind that you are the best at the job you can be. A huge part of "survival mentality" is your commitment to training, education, and personal preparation.

PROPER PRIOR PLANNING PREVENTS POOR PERFORMANCE

Amateurs train until they get it right; professionals train until they can't get it wrong!

In order to face things that may come your way tomorrow, you need to begin planning today.

With the proper tools and knowledge we can do anything we put our minds to. Recruit school taught you lots of practical, useful knowledge and skills. Recruit school did not in most cases make you excellent in all of the skills needed to survive. It is all about preparedness and a good attitude. You will face a lot of naysayers as you work on your personal preparedness to be the best. Don't let them sway you from your "survival mentality."

We can often improvise even without the right tools for the job if we have the knowledge and inspiration to do so. It's all about preparedness and a good attitude. If we believe that we can overcome whatever it is that comes our way, we will be able to do so. This is what I am talking about when I use the term "survival mentality."

This preparation will cost you something—time, money, or whatever else is necessary. If you are going to become the best you can be, you must be willing to deny yourself today for the possibility of survival tomorrow. The price will be small in comparison to the later rewards. Are you willing to make the sacrifices? It is solely your decision and commitment.

If you will prepare now, it will pay off in the long run, even if you never have to face a survival situation. The Marines adeptly puts it this way: "Improvise, Adapt, and Overcome." Begin planning today so that you can conquer tomorrow!

EMBRACE THE POWER OF POSITIVE THINKING

Make a conscious decision to look for the positive aspects of your situation instead of wallowing in the negative responses common to those who encounter life-threatening situations, including anxiety, fear, anger, boredom, frustration, loneliness, or even suicidal thoughts. Adopt a hopeful and expectant attitude, which will enable you to remain open to solutions that you might not have considered. Maintaining a positive outlook will also keep you open to creative and innovative approaches that are less likely appear in moments of despair.

BE REALISTIC

Finally, be honest about your situation. But instead of letting a grim appraisal get you down, use the truth of your predicament to motivate you to be your best, both physically and mentally. Hope and expectation won't get you anywhere alone, so keep working toward your survival plan until you reach safety.

SURVIVAL MENTALITY TIPS

- 20% of your skills will be used 80% of the time, *but* the skills you need during the remaining 20% will more than likely make the difference between your life and death.
- Do not be the firefighter, officer, or chief who has quit but stays on the job. When you are no longer enjoying making a difference with the people you work with, retire or quit.
- Firefighters stay longer and are more engaged if leaders demonstrate appreciation.
- Do not reject "troublemakers"— those who are uncomfortable with the status quo. Sometimes frustration or anger can result in innovation. There's a fine line between the complainer and the leader aiming to fix things that bug him.
- Perception is everything! Facts do not always matter in how things progress. *Kindness is free.* Listening to firefighters or patients and answering their questions costs nothing. *Perception is all there is—* you must become a master student of all things associated with the creation of perception surrounding the provision of any and every product and service.
- Life is theater! All the world is a stage. Every one of us is an actor (100% of the time). "Getting" the "theater bit" is the essence of strategy. Acknowledging "theater" as the centerpiece of effectiveness in implementation and acting accordingly is of the utmost importance.
- Study all of the time. Take the fire business seriously and remember, "what you don't know *will* kill."

- Call a friend. Right now. (Stop reading this—make the call now!)
- Become a student of all you will meet with. The business of being effective leaders is first and foremost about relationships.
- Training is not time wasted but should be time made for and valued. Training, knowledge, and experience make you a safer firefighter.
- Learn from other's experiences. Call other firefighters, officers, and departments to learn from them when they have a fire.
- Become a student of yourself. Being aware of yourself and how you affect everyone around you is what distinguishes a superior leader.
- Hang out with interesting new people. Manage the "Hang Out Axiom" as if it were a life or death issue—it is. Hang out with "interesting"—get more interesting. Hang out with "dull"—get more dull. Little, if anything, is more important for innovation than precise "hang out management." Measure it!

Chief John M. Buckman III

Dr. Harry Carter

Holy Crap—I Almost Died!

I first met Harry after seeing an article he wrote about himself getting trapped in a fire. Not in the urban fireground where he worked, but in his hometown, where he served as a volunteer firefighter. It's a striking photo that—if it's even possible—allows the viewer to somewhat feel what that scene was like. I know when I saw the pic—it worked for me. Unfortunately, that photo couldn't be found, but imagine a soot covered, shaken, disheveled holy-crap-I-almost-just-died look on his face, and that's a start for you to imagine what he went through. It was real close.

Harry is a veteran chief fire officer, instructor, author, writer, commissioner, and lecturer. He also likes to chat—and with his decades of experience, it's always interesting! He teaches at a number of community colleges as well as the National Fire Academy in Emmitsburg, Maryland. He retired in 1999 as a Suppression Division battalion commander with the Newark, NJ, FD. In Newark, he served as chief of training, commander of the hazmat team, and assistant to the fire chief. He is also a past fire chief and former training officer of the Adelphia Fire Company, NJ. He currently serves as the fire company chaplain and active apparatus driver. He's written thousands of pieces in the style he speaks: lively, animated, and a mix of old school and modern "get it." Harry has always passed it on.

Dr. Harry Carter with his firefighter daughter Katie

Avoiding the New Boss Blues

For many years I have had my own personal gripe with the fire service. I have long complained that the tacit knowledge (not book learning but experience) acquired by veterans during their years of service was not being shared with the younger members who will be around long after we old dogs have stopped learning tricks of any kind. If we are to broaden the knowledge level of the fire service and create a safer environment, we need to improve the concept of sharing what you have learned as you have gone up the ladder of command.

The jump from firefighter to company officer is an important part of a person's career in the fire service. Let me begin by stating the obvious. Far too often the only thing you receive from your fellow travelers is a word or two of congratulations. Think about the significance of this move.

You have been placed in a new position of responsibility within your fire department. Maybe you have undergone an extensive period of studying, promotional examinations, and personal interviews. Then again, you may have been elected by your friends to this new position. There are also those among you who may periodically end up in the right front seat in an acting capacity.

You are probably feeling pretty good about yourself right about now. I know that any promotion I ever received was a time for good cheer and celebration. So spend a bit of time enjoying yourself. There is nothing wrong with reflecting upon your success. Take the time. You have earned it.

However my friends, no matter how you got to the point where you are today, be ready for the letdown. There is always a letdown of some sort after each moment of joy. So it has been throughout my career in the fire service. I can recall my first time in the right front seat in the U.S. Air Force. It happened to be in the spring of 1968.

I had spent a great deal of time studying and felt that I was ready for the new job and its attendant duties and responsibilities. That made one of us who felt I was ready. I was assigned as a crew chief on a crash-rescue vehicle at the crash station on the flight line at Eielson Air Force Base outside of Fairbanks, Alaska. Today you would call me an ARFF guy.

My problems came not from the close inner circle of friends I had nurtured during my time as a driver and firefighter. Most of us had attended training school together at Chanute Air Force Base in Illinois. These folks knew me quite well and knew that I would have their interests at heart. No my friends, the problems came from other places.

There were those who felt that they had been cheated by the system. They felt that since they had been in the service longer than me, they should have been promoted ahead of me. They took their frustrations out on me. What they failed to note was the number of things my associate and I had done to prepare ourselves for leadership roles.

I had a number of college credits on my record and worked hard to do the right things. Correspondence classes, previous job experience, and a great deal of extra work within the fire department allowed me to grow as an individual and impress my superiors with the quality of my work and the level of my dedication. So too it was with my buddies who were promoted at the same time as me.

My roommate and I spent a great deal of time working with our fire prevention division. Each month he would don the costume of Sparky the Fire Dog, while I wear the Smokey the Bear suit for our public education effort. It was fun for us and had a strong positive impact upon the dependent children at Eielson Air Force Base. We went about the call of duty.

So I felt qualified for the responsibilities I was given. That did not make it any easier to endure the snide remarks and backstabbing actions of others. There were a few of us in the same boat, and I guess we looked to each other for the strength to endure the onslaught of negativity that we faced.

There were those who tried to play the cards of oppression and prejudice. They stated that a number of us were promoted because of our race. I discovered early on that one can never control what other people will think. You will also learn early on that not everyone is a supporter and encourager of your career. They have plans for themselves, and you are not on their radar screen.

These negative vibes have gotten worse as our society moved through the "me" generation to the current model that I have come to call the "what's-in-it-for-me" generation. I promise you that you will hit the brick wall of peer jealousy very early on in your new role as a rider in the right front seat. This holds true whether you are in a career or volunteer fire organization.

A few years later, I faced a similar problem in the Newark, New Jersey, Fire Department. The negativity was the same, but the reasons were different. I was promoted to the rank of captain after only four years and one week of service.

I was only 30 years old. Many of the guys who got promoted at the same time as me had at least double the amount of service.

Jealousy is not a pretty thing to see. Sometimes it is blatant and sometimes it is subtle. However, I would like to caution you that it will be at work in and around you as you move into your new position. Be on the lookout for people who will stab you in the back at their earliest opportunity.

The key to both of the situations I have mentioned comes from knowing what will happen and then making up your mind that you are tough enough to leap the hurdles that will be placed in your path. You must also be ready to do battle with older people of the same rank.

I can remember having a real pitched verbal battle with a fellow captain who insisted on calling me "kid." This problem came to a head one night after a particularly bad fire when this guy's company stretched a hoseline off of my pumper company, and after the fire left the hose in the street for my crew to reload. My crew was tired of the actions of this man and his crew of prima donnas.

They had pulled this nonsense on us before and this night just happened to be the straw that broke the camel's back. When I approached him about it, he tried to provoke me into a fistfight at the fire. Fortunately I was not stupid enough to succumb to his challenge. Hell, he probably would have kicked my butt anyway.

No, I suggested to him that he and his truckload of prima donnas return with us to my station for a further discussion of this issue. I knew that I was going to win when he agreed to my suggestion. When we got to the quarters of Engine Company #15, I asked him to come upstairs to my office. If anything negative were to happen, let it happen behind closed doors was my thought.

The interaction was quite brief. Before he could utter word one, I stared him down and told him that I wanted him to cut out the "kid" crap. I told him that as a veteran of the Vietnam Conflict, I had long since passed the "kid" stage of life. I further told him that I earned my rank in the same way that he did. We had both studied and taken the same sort of civil service promotional examinations offered by the state.

I looked him right in the eye and indicated that I wanted us to be friends. I extended my hand to him and after what seemed like a really long time, he un-balled his fists and grudgingly extended his hand toward mine. I do not think that we ever really became buddies. However, we did develop a healthy respect for each other. They reloaded our hose in the future when they stretched it at a job. And I did not end up getting my butt kicked.

The key to the new-boss conflicts is keeping your cool and taking the moral high ground. The same holds true for the interaction between you and your crew. Many times you will end up with a crew that is more senior than you. That is how it was for me in Newark. Most of the guys on my first crew were old enough to be my father.

If you want to win people like this over, you need to do it with your interpersonal skills, your technical skills, your honesty, and your sincerity. You need to learn as much as you can about who your people are and what motivates them to be team players. You must then create a team based upon the skills and interest of both you and your crew.

A buddy of mine taught me the value of the team meeting. Much can be accomplished by the honest discussion of the problems facing your team. It is critical to lay out your plan for running your company. You must also make sure that what you want to do is allowed by your department.

The person in the right front seat has to become a font of knowledge as regards your department's rules and regulations, standard operating procedures, and operational guidelines. You must act within the guidelines established by your department.

If you establish yourself as one with little regard for the rules, do not be surprised when your troops begin to challenge you. If you disregard the rules, you will engender an example that will come back to bite you on the butt. This is critical.

I am sure that I will share a few more bits of wisdom with you as the months pass and we move further on down the road to helping you become a better officer. Rome was not built in a day. The same is true with regard to your career as an officer. Let me close by saying that no one will ever become perfect.

However, if you fail to work at self-improvement, I can guarantee that you will travel a long road as you sit in the right front seat of your fire truck. It isn't easy and it doesn't happen without effort on your part. No one just becomes a leader. Each of us who has been granted the privilege of leading others has had to learn how to do the job.

So it is with you. You are nothing special. Yet at the same time you are very special. You have the ability to influence the manner in which people work. If you are good they will be good. If you are not good, they will not be good. In either case the people in you world will see what you are doing. You cannot hide from the truth. But be sure to share it.

Sal Cassano

A Fire Commissioner Who Does It... From the Heart

Sal Cassano is absolutely one of those commissioners/chiefs who "gets" that he is responsible for every EMT, paramedic, and firefighter under his command, and he takes it very, very personally. A son of Italian-American immigrants, Sal was born in Brooklyn and lives on Staten Island. After serving in the U.S. Army during the Vietnam War, he began his FDNY career as a firefighter in November 1969. He has served in all five boroughs of New York City and has received five commendations for meritorious acts for bravey. Commissioner Cassano, who has a bachelor of science degree in fire science from the John Jay College of Criminal Justice, has held every rank in the FDNY (from firefighter to chief of department to fire commissioner) and was instrumental in rebuilding the department after the September 11 attacks, after which he was named chief of operations. In 2006 he was appointed chief of department, the highest uniformed position in the department, and was appointed as fire commissioner in December 2009. In 2009, he received France's highest award, the Legion of Honor. In 2011, in recognition of his outstanding service to the city of New York, he was awarded the very prestigious Ellis Island Medal of Honor. For those of you who are runners, Sal is a marathon runner who has run the New York City Marathon four times and has broken the three-hour mark. Below are some very personal words from Sal Cassano, a leader who has set the example operationally, physically, educationally, heroically, and most importantly... all from the heart.

A Strengthened Sense of Purpose

When Billy asked me to contribute to this book, it was an honor considering my deep respect for the National Fallen Firefighters Foundation and of course, for my late friend, Chief Ray Downey, who was killed along with 342 members of the FDNY on September 11, 2001.

On that day the lives of every member of the FDNY, both uniform and civilian, changed forever. Our city was attacked like never before. Unimaginable devastation and unthinkable loss were brought to our doorstep, and we were faced with the awesome task of rebuilding our department. We owed it to all 343 members killed that day to not let their deaths be in vain. We never forget them and we will always honor them for their sacrifice. As anyone who has ever served in a fire department knows, these were not simply people we worked with; they were our family. I lost my closest friend, Chief Gerry Barbara; my mentor, First Deputy Commissioner William Feehan; and two men who taught me more than I could ever thank them for, Chief Peter Ganci and Chief Donald Burns. The names and faces of all 343 members are visible throughout the department, ensuring their memory can never fade. Their sacrifice is what drives us to be better every single day.

September 11th was a day when extreme violence, horror, and fear were unleashed on thousands of innocent people. But that is not the whole story, because on that terrible day there were moments of incredible loyalty, kindness, and selflessness. An evil attack that was brought about by a lack of humanity resulted in extraordinary acts of courage and devotion to others. No group embodies that devotion and service as much as the New York City Fire Department. As commissioner, you would expect me to feel that way—and I do—with every beat of my heart, but that kind of pride should not be limited to just FDNY members; it should be felt by every firefighter in every department that they have been honored to serve in. That day, the world saw tragedy, but they also saw the remarkable actions that firefighters will take to save a life.

All firefighters in every department have been on the front lines for their community, just as we were that day. At the World Trade Center, we fought fire, evil, and hate, not just with physical strength, but with courage, commitment, and compassion. In the immediate aftermath, we were battered and dazed. We knew that things were bad for us. We knew we had lost many friends. We knew many families were scared and hurting. But just a few days after the attacks, while working at the site, I came across a firefighter I knew who made me realize just how bad it really was.

He was walking around and was clearly upset. I asked him if he was looking for the members of his firehouse who were missing. He looked at me and said: "No, Chief, I'm looking for my sister." She just happened to be at the Trade Center that morning for a meeting when the planes hit. She, too, was killed. That really brought home to me that it wasn't just those of us who had trained for danger who had been caught in this terrible act. It was many innocent people who were simply going about their business, making a living for their families. It was then that the magnitude of the loss—not just for us, but for the city and even the country—really began to sink in for me. The department had suffered catastrophic losses, and the families of our members who survived were also experiencing tremendous loss. At that moment, I realized that the victims of the attacks and those who loved them formed a community far wider than just the fire department. That community exists to this day.

Since that terrible day, we have recommitted ourselves to devotion and service with a new and strengthened sense of purpose, not only because our mission is to protect the citizens of this city, but because it is the best way we know to honor those we lost at the World Trade Center.

We were able to succeed because we had the memory of the 343 with us at every moment, giving us hope when we thought all hope was lost and giving us strength when we thought that we just couldn't go on. But the damage done when those towers came down continues. Since that day, we have seen many of our members who worked at the World Trade Center develop illnesses directly related to their tireless and selfless work to bring home those we lost. Once healthy and strong, these men and women have succumbed to terrible illnesses. Far too many of our families have experienced this loss, and unfortunately, the list continues to grow. Those fallen members are casualties of the fight that began on September 11th.

Finding the strength to go on after a catastrophic loss is difficult. Every one of us at FDNY had a choice after that day. We could give up, turn around, and go home; or we could stay, band together, and rebuild. For every member, the choice was clear. The world witnessed the FDNY come together as never before. While our firefighters, EMTs and paramedics continued their outstanding work in the field; they also continued to search for those lost. The civilian staff in the department rose to the occasion as well, pulling together to provide those field members with the support they needed to continue to get the job done.

The amount of effort by our civilian staff here and elsewhere in this department before that first-due engine and truck can go out the door, and before that first ambulance can be dispatched, is inspiring. Don't forget those important people at each of your own departments. Without them, we, as firefighters cannot succeed. They are an integral part of our department, and yours as well.

Sal Cassano in 2000 at a multiple alarm in the Bronx

Like you do for your own department, I have a deep sense of pride in mine. I feel it's the best fire department in the world, and not because we're the biggest and not because we are the busiest—it's because of the caliber of people I have had the pleasure of working with for the last 44 years, from the civilians to the members in the field. It's because 12 years after the worst attack we have ever experienced, we stand better equipped, better trained, and better prepared than at any point in our history. We have rebuilt and shown a resiliency we could have hoped for, but never imagined. The members of the FDNY have shown that no matter what challenges come our way, we can never, and will never, be broken. We made it our mission to learn as much as we could from our horrific losses to honor those who were lost. And it ensures that we will never forget.

So how does this impact you, as a firefighter but not a member of the FDNY? First and most importantly, especially those who were firefighters in 2001, we know you were there for us and supported us in so many ways. We are forever indebted to all of the firefighters, departments, and first responders from across the country and around the world who stood by us during our darkest hours. No community supports its fellow members like the fire service. I know that still today, thousands of your departments hold memorial services and have erected permanent memorials honoring not only your fallen firefighters, but also all those who were lost on September 11th. We can never thank you enough.

Perhaps, as my way to thank you, I can offer some thoughts that apply to firefighters, marshals, officers, and chiefs, to remind you how fortunate we all are to do this job.

We firefighters are different. Just ask our friends and families! We put our lives on the line for someone we've never met and for the person sitting next to us on the rig. We have chosen a profession, be it paid or volunteer, defined by honor, bravery, and service. It's a tremendous responsibility and, therefore, we are held to a higher standard both on and off the job.

Since I first came on the job in 1969, the duties and responsibilities of firefighters have expanded far beyond the scope of previous generations. Today's firefighter doesn't just respond to fires. In a moment's notice, firefighters at any department can find themselves responding to medical emergencies, incidents involving hazardous materials, natural disasters, and potential acts of terror. This kind of dynamic environment demands that you constantly learn, adapt, and stay current in an ever-changing world. But despite these changes, one thing has remained constant—the firefighters themselves. Though the job has evolved, it always requires a different type of person from the average citizen.

I remember my first day as a probie. Walking into the firehouse for the first time is one of those moments you never forget. It's like that for every firefighter. While the individual experiences may differ from one department to the next, I believe there are many common threads: leaving early each morning for the academy and coming home exhausted and sore from training and smelling of smoke, falling asleep with your face buried in a book studying every night, and gaining the respect and support of your family and friends, just to name a few. When you entered the fire academy, odds are someone told you that it would be the most difficult weeks of your life and that your instructors would push you like never before. And every firefighter reading this book knows they were right.

I hope your instructors pushed you to your limit. I hope they made you stronger, and they made sure you left the academy in the best shape of your life—because your life and the lives of your fellow firefighters depend on it. Like our firefighters in the FDNY, you were instructed in the history, customs, and traditions of your department. They introduced you to the apparatus, tools, and equipment of our job and instructed and drilled you relentlessly in the firefighting tactics and procedures that are the stock and trade of our everyday existence. These lessons must become second nature to you in order to effectively and safely perform the noble mission with which we are charged. You fully learned that being a firefighter means being part of a team. The time in the academy is when you saw first-hand that true commitment and support to each other is crucial for the job we do.

For those of you about to enter the academy, get ready for the time of you life! It will be difficult, but the rewards of this job make it all worthwhile.

When people ask me about the kind of person it takes to be a firefighter, the most recent FDNY probie class in May 2013 comes to mind. One story from that class really shows the character of the class and demonstrates what this job is all about.

Prior to the start of training, one member of the class had his home destroyed by Hurricane Sandy. Despite this personal setback, his desire to be a New York City firefighter was so strong; he wouldn't let anything deter him. He attended probie school by day and worked to repair his home at night, while his wife and child were forced to live elsewhere. When they learned what he was going through, his class and the instructors at the academy stepped up to help him with supplies to rebuild his home and worked there on the weekends to help get things back in order. Already, the commitment to each other was taking place, before any of them had set foot in a firehouse. This member was deployed to the military before he could complete probie school, but will be back in the academy when he returns home.

In the days after graduation, those probies reported to their assigned firehouses, but another probie was not one of them. He had been called to active duty in the Navy and left to serve his country, putting his dream of being a New York City firefighter on hold, for a short time, to fulfill his service to his country. To go through 18 weeks of probie school and then step away from it all to defend this country demonstrates remarkable dedication, honor, and bravery. To me, that story demonstrates the outstanding caliber of people we are so blessed to have in the FDNY. That's the kind of person who becomes a New York City firefighter. Those are the values and the commitment that separates this job from so many others.

It takes a special person to do this job anywhere, as every member of every department knows. Running into burning buildings, responding to natural disasters and terrorist incidents, and risking your life for another are incredible feats. Firefighters do them all every single day.

For the newest members of the fire service reading this, always remember that your education did not end when you graduated the academy. Know that every day you go to work you will be learning, and that education continues off-duty as well. I've spent 44 years in this department, and I'm still learning, so you all have a long way to go! If you don't understand something, ask questions. If you don't know what to do in a situation, ask questions. If you don't know why you are doing something, ask questions. Don't try to improvise on the way you were taught and trained to do things. This is a job that gets very real very quickly. In this business, sometimes we can get lucky, but luck isn't enough. Those who taught you and will continue to teach you have *been there and done that.* Always respect their experience. Sometimes we're heroic, but heroism isn't enough. In our profession, knowledge is power, just as it is in other endeavors. Never stop learning.

Firefighting is a team effort. We're only able to do this job so well because of the firefighting team around us. Our efforts to protect the public are tremendous: the effort to protect each other is nothing short of extraordinary. You should know this to be an absolute certainty. Firefighting isn't in your blood; it is in your heart, and those around you will quickly read your heart, through your actions.

Embrace the history and the traditions of your firehouse and company, ask questions, and learn as much as you can about the work you do. The firefighters in your company want to make sure you succeed. Every firefighter, officer, and chief wants to make the next generation that much better, so that the traditions of bravery, commitment, and sacrifice we hold so dear live on.

Perhaps some of you are new officers, studying to be promoted or have been officers for some time. The transition from firefighter to fire officer can be very difficult, but also very satisfying. It's the hardest adjustment you'll ever have to make, so do so cautiously. As a fire officer, you don't follow, you lead. You set the tone for the group, and the firefighters look to you for direction. You must ensure that the members operate safely, wear their protective equipment properly, and always respond in a timely manner. You need to conduct meaningful drills, and critique their operations to ensure they are always improving. The decisions you make have a profound impact on the safety of your members and the public. Your duty is to ensure that your units return safely from all operations. To help you carry out this mission, you're given the apparatus, tools, equipment, and procedures needed. But none of that matters unless you are completely committed to your firefighters.

This strong leadership requires a tremendous effort on your part as well as your department. In the FDNY, from lieutenant to chief of department, our most important mission is to protect our members from the dangers of the job, whether it's the ravages of fire, or from the challenges of dealing with medical and other emergencies.

Some of the toughest decisions you have to make are when you have to say no. I tell every group of newly promoted officers in the FDNY, "If it doesn't feel right, it probably isn't." And while you do have a chain of command, remember that you also report to the families of those under your command. They expect you to do the right thing for their loved ones.

Perhaps some of you reading this are involved with fire prevention, investigations, and code enforcement. You are in one of the most underappreciated bureaus in any department. The work you do and the manner in which you handle fire investigations results in a safer work environment for your members and the public. When an arsonist is removed from the streets, their depraved indifference to life goes with it. With the prosecution of an unscrupulous landlord comes the assurance that illegal construction or renovations will not maim or kill our members, citizens, or visitors to your community. Remember that your work is extremely important, and every member of your department and every member of your community counts on you.

Some of you may be new or relatively new chief officers. You are the key link between the top leadership of your department and the field when it comes to communicating the directives of the department and the needs of the field units. You're close enough to the members to know what they think the department should be doing for them. At the same time, your prominent role means you are charged with passing the philosophy and directions to the officers and firefighters under your command. You don't have an easy job, but it is very gratifying when you see the fruits of your labor.

I hope that by reading this you understand the pride I have for my department, the FDNY. I challenge you to share in that pride by doing your best to improve yourself as a firefighter, officer, or chief and to feel that same sense of pride in your own department. Set an example for teamwork, trust, adaptation, discipline, and respect. Enjoy your days as a firefighter, because before you know it, this too shall pass.

Let us remember the many firefighters who have given their lives in the line of duty across North America and the world. Please remember the 343 FDNY members we lost—ranging in rank from the chief of department to the newest probie, who took part in the greatest rescue operation the world has ever seen.

May God Bless the memories of those we have all lost as firefighters, and may God continue to bless your department and the members of mine. And to borrow some of the words that my good friend, former chief of department, the late Peter Ganci would say . . .

"Twenty or thirty years from now, when someone asks what you did for a living, you can tell them you were a firefighter . . . and it doesn't get any better than that."

James Clack

A Tall Man with a Clear View for Firefighter and Civilian Survival

I got to know Jim when he came to Baltimore after more than two decades of service in the Minneapolis Fire Department. I guess what really drew my attention was his genuine interest in connecting with the people, but more specifically, the members of the Baltimore FD . . . on social media. I thought it took a pretty tough and open chief coming in from the outside to engage "openly" with the members in chat groups and related rant sites.

When then Minneapolis Fire Chief Jim Clack arrived on the scene of the infamous I-35W bridge collapse, he had a hard time believing it was real. I think after getting a taste of Baltimore politics, he may have had similar feelings.

That was the beginning of his journey of change to a department that had been through some rough times—including the death of a cadet firefighter. To me, it seemed Jim's blatant focus was the safety and survival of his personnel—and his citizens. I never saw or read where he tried to tell them to fight fires "differently," as their record of saving lives was well known. But moreover, his interest and leadership was focused on reducing death—all kinds—and he did that. Under his leadership, the department reduced fire deaths to the lowest level on record. How?

In 2010, the city passed his legislation to require installation of sprinkler systems in all newly constructed one- and two-family homes. He also secured federal and private funding to support the installation of smoke alarms with 10-year lithium batteries in homes at no cost to the residents. It worked. Saving lives at fires isn't always about heroics on the scene, sometimes it's about political and legislative heroics. He took his experiences and passed them on—saving live along the way.

What Happened to Us?

Back in the day, we signed up to be Firefighters, not EMTs or fire inspectors. These were day jobs for people who could no longer fight fires. Most of us wanted only to battle fire with a hose and an axe. If we were lucky, we might save a life once in a career by dragging a nearly dead customer or two outside. This was exhilarating work with good benefits and lots of time off, and most everyone in the community loved us.

Many fire officers of past generations did not place any emphasis on formal education and training. This was a "learn on the job" (OTJ) profession; there was no need for fancy book knowledge. Just do what the old guys tell you and you will stay alive and maybe earn the respect of your coworkers before you retired.

What the hell happened to our profession? EMS response and fire prevention activities have become accepted by most as primary roles for the new fire service. Fire deaths and structure fires in general continue on a long-term decline. Our equipment is much better than even 20 years ago: our stuff is lighter, stronger, and just plain better. One could argue that these are all good developments.

But something else has changed: Our public no longer accepts what we do and how we do it without asking some questions. Questions about our schedule, our benefits, even how we go about doing our jobs. Many people now pointedly ask who we are and why we deserve to be held in such high esteem. This questioning can quickly turn into accusations that are very public indeed, accusations that hit us at the speed of light on the Internet in the form of blogs, comments, and conversation boards. This change in public perception has real implications for us. We must evolve and do it quickly.

The ability of the fire service to maintain real significance in the hearts and minds of the public we serve will depend primarily on individuals. Can we become much more nimble, adaptable, and open to change? Our future leaders must commit themselves to the hard work of developing individual competencies in various disciplines outside of the fire service. This knowledge will then need to be translated into new public services that are valued in the community. Promotions from within the organization will become promotions from anywhere those in charge can find relevant knowledge and leadership ability. There will be an increasing need in the fire service for "generalists" as opposed to "specialists." Individual personal development is a wise personal and organizational strategy to be ready to meet tomorrow's changing landscape.

That is the "lens" by which what we do will be judged by the public. A successful fire service career will be directly related to each individual leader's ability to absorb new knowledge, learn new skills, and develop effective interpersonal relationships with other people from all walks of life. Our survival as a profession depends on it.

Chief James Clack

Ice house fishing in northern Minnesota

Burton A. Clark

Higher Education from an Urban Firefighter

Dr. Burton A Clark, EFO, was one of my instructors during my five years at the National Fire Academy Executive Fire Officer (EFO) Program. Fueled by passion and real-world experience, he has become well known in the last few years for his angst over unnecessary firefighter deaths. You may be unaware that, while a professional educator, Burt has been in the fire service for 40 years, was a firefighter in Washington, D.C., and an assistant fire chief in Laurel, Prince Georges County, Maryland. He is the Management Science Program chair in Emmitsburg and serves as an operations chief during national disasters and emergencies at the national level. He has also been a visiting scholar at Johns Hopkins University since 2011, researching fire service safety culture. Burt is one of those who is able to strike a nerve and get you to think . . . think hard about how we do business, both administratively as well as operationally. Because he makes us pause and think, the job is better because of Burt's efforts.

A firefighter's gear works best when there is a strong education underneath it.

What You Do Wearing the Helmet Is Only as Good as What You Put in Your Mind

In 1868, Sir Eyre Massey Shaw, fire chief of the London Fire Brigade wrote, "When I was last in America, it struck me very forcibly that, although most of the chiefs were intelligent and zealous in their work, not one that I met made even a pretension to the kind of professional knowledge with I consider so essential. Indeed one went as far as to say that the only way to learn the business of a fireman was to go to fires—a statement about as monstrous and contrary to reason as if he had said that the only way to become a surgeon would be to commence cutting off limbs without any knowledge of anatomy or of the implements required."

In 1976, I wrote, "If there is one thing common to all fire departments in the world, it must be bull sessions between emergencies. During a recent session, I was defending the importance of an academic education for the firefighter. Needless to say, I was alone in my opinion. The majority argued, "You don't need a college degree to ride the back step." I countered with, "If we become more educated, maybe we'll kill fewer firefighters." My lieutenant said, "Firefighters have to get killed; it's part of the job." I have rejected that notion for more than 40 years. When there is a firefighter injury or death, something went wrong; it is not part of the job.

In 2012, I told a recruit firefighter class, "Each one of you will have to develop your own yardstick over the next 30 years to measure satisfaction. What will give you fulfillment and gratification? What will meet your needs, desires, and appetite? How will that change over time? The fire service discipline is truly a calling, which offers unlimited opportunity to test you mentally, physically, and emotionally. I cannot give you a measuring device, but I will share with you four ideas that have helped guide me over the past 40 years, continue to help me every day, and shape my future directions.

"First, the art and science of being a firefighter is a life and death vocation that requires a 200% proficiency level. You must perform every task 100% correct, 100% of the time. Because over the next 30 years, you do not know which task, at what time, could be a mistake that causes an error, leading to a failure, which results in a tragedy.

"Second, society trusts firefighters. It is your duty to act professional, follow a code of ethics, and deliver the highest standard of care in every act you perform. What will help you reach this goal is to treat all people as your family, friends, and neighbors.

"Third, do your best all the time. Then determine how you can do better the next time. This will require life-long education, training, practice, and research.

"Fourth, honor the past. A lot of people and events got you to where you are today. Celebrate the precious present. You have accomplished much and tomorrow belongs to no one. Believe in the future, you are responsible for creating it."

Finally, for all present and future firefighters, remember: As a member of the fire service discipline, what I know and what I do not know has life and death consequences for me and others. Therefore, a fire service calling demands my highest level of professionalism.

It is my honor to share these thoughts with you.

Burton A. Clark

First Fire—Important Lessons

August 1970, a Saturday afternoon in Landover, Maryland; I had just moved there from New Jersey and did not know anyone. My wife sent me up the street for bread. When I came out of the store, I noticed the Kentland Volunteer Fire Department Co. 33 across the street; the doors were open. No one was there except the Dalmatian named Halligan.

A couple minutes passed, the fire trucks returned, the firemen were in dusty softball uniforms. We talked, they handed me a beer and asked if I liked to play softball; I said sure. So, I became a volunteer fireman on the spot.

I had been in the fire department for about two weeks but never actually got on the fire truck. One afternoon during the middle of the week I was at the station sitting in the TV room watching *The Gong Show*. A young guy in an Army uniform came in the room. We introduced each other; he was Jimmy Panetta, a member who just came home on leave. We talked; Jimmy asked if anyone had shown me anything. No, they just told me to get gear off the spare rack and get on the truck. Jim said, "Let me show you some stuff," and we went to the apparatus bay.

He proceeded to show me how to stand on the back step of the pumper—hold on to the bar and bend your knees. Then he said he was going to explain how to be the layout man. He showed me the bundle of hose with a rope around it. He said, "The truck will stop at a fire hydrant; the officer will open the door and yell out 'Layout.'" I was to step off the back step, pull the rope and hose off the truck, wrap it around the fire hydrant, and yell "Go" the to the driver. I got it—now I am a real fireman. Back to *The Gong Show* we went.

Ten minutes later, the bells and lights flashed and rang, we got something. I put my gear on and got on the back step. Four of us responded; the paid driver, the paid sergeant, Jimmy in the bucket, and me on the back step. We turned left out of the firehouse, went four blocks, and turned right into my apartment complex. Fire was blowing out of the third floor windows.

The engine stopped, the officer opened his door, and yelled "Layout!"; there was a hydrant on the left, so I stepped off, wrapped the hose around the hydrant, yelled "Go," and I was *done*. I had no idea what to do next. I just stood next to the hydrant, watching the fire. It seemed like forever for help to arrive, but within two minutes the next pumper from Kentland pulled up with only a driver, Wayne Ramsey. We don't know each other, but Wayne did know that I had no idea what I was doing. The steamer cap was not removed from the hydrant, the bundle of hose was still warped around the hydrant, and hydrant wrench was stilled hooked to the rope.

The hydrant was on a grassy island in the parking lot so the pumper could nose right in, only 4 or 5 feet from the hydrant. Wayne knelt down to untangle the hose and take the steamer cap off. He told me "Get that big hose off the front bumper and bring it to me." I could follow directions. The hose was the big 5-inch soft sleeve with the 15-pound brass coupling on the end of it. I grabbed the hose about 8 inches behind the coupling, pulled it, and swung around to give it to Wayne. The coupling hit him in the side of the head, and he went down like a rock. I realized that not only did I not know what to do, I was dangerous.

Wayne shook off the clubbing and said, "Stand there and don't move." For an hour I was frozen in place watching this amazing ballet of firefighters. The longer I stood there, the sicker and sicker I got in my gut; it was a terrible feeling, one I had never experience before, and I never wanted to feel that way again. I promised myself I would never not know what to do next. There was more to being a firefighter than softball, beer, and *The Gong Show*.

November 1972, I graduated number 1 from the District of Columbia Fire Department Recruit Class 249. Today, I am still learning what to do next.

Hank Clemmensen

Leading the Chiefs

As of this writing, Hank Clemensen is the president of the IAFC. A firefighter since 1972, he brings decades of experience to the association—representing the leadership of the fire service. His cool, calm, and see-through-the-fog focus has provided the IAFC with a solid leader in some really difficult times. He served as a firefighter, paramedic, and fire marshal. He was president of Illinois MABAS Division 1, president—IAFC Great Lakes Division from 2007–2009, and vice president of the IAFC from 2010–2012, and he is also a commercial pilot/flight instructor for airplanes and rotorcraft-helicopters. Fighting fires, pushing fire codes, being a paramedic and a skilled pilot and flight instructor? Sounds like the perfect skills for the president of an association challenged with leading fire chiefs!

Dedicated to my father, Hank Clemmensen, Sr.

Acceptable Risks

What I believe today is that every firefighter, paid, volunteer, or POC, should be required to attend the National Fallen Firefighters Foundation (NFFF) memorial weekend in Emmitsburg, Maryland, early in their career. I would then challenge every firefighter attending the memorial to walk up to the family of a fallen firefighter, whether it's the parents, wife, children, or siblings, and thank them for the sacrifice made by their loved one. I only wish you could tell them that every single LODD has made a true difference in saving another life. It would be my hope that no family has to come to terms with knowing that their loved one died because they were simply driving too fast, trying to save a vacant building, or just doing something irresponsible that was against department policy or procedures.

This could be the worst time ever to be a fire chief. With tax revenues down and our budgets being cut, departments are being asked to do more with fewer firefighters and less companies. As the fire chief, some of those little things we read about or heard about at the last leadership conference might seem simple, but they aren't easy. The fire service can be like a pressure cooker of stress, with the everyday administrative issues interrupted by moments of sheer terror from emergency responses. Yet the best leaders stay calm despite all that is going on around them. When it would be easy to yell or scream, the true leader relies on their experience and puts that stressful situation into perspective.

When I first got hired as a full time firefighter in the mid-seventies, the department was made up of a lot of old school firefighters and officers. Today, that same department is made up of the newer breed of firefighters, but they still have those same old-school traditions and culture that have been passed down for generations.

What amazes me is the courage and complete disregard for our own life and safety when it comes to saving other people's stuff. As George Carlin would say, "It is all stuff . . . and a house is just a place to keep your stuff while you go out and get more stuff." This is the same person who said, "Life is worth losing," and now he is dead.

The bottom line is that the business of fighting fires needs to change, like it or not. So why can't we change our methods of firefighting to meet these new challenges? Why is it that we can't understand that an interior aggressive attack is not necessary when there is no human life to save or at risk? Why can't we change how fast we drive to those fire alarms that we know have a 99.9% chance of not warranting an emergency response? Why is it that once we turn on those emergency lights and sirens, we drive faster and can't seem to stop at red lights or stop signs? Why do we fight a dumpster fire as if it were full of people rather than just garbage? Why do we fight fully involved car fires as if we really think we're going to save something? Why are we so surprised that the facepiece on our SCBA melts at 500°F, and do we really need it to withstand higher temperatures? If the people we are attempting to save can't survive 500°F, why in the world are we in those types of temperatures?

We must accept that the world is constantly changing around us, and it's time for some of those 100-year-old traditions to be honored as part of our history. It is this culture of the American fire service that is killing our firefighters today. Changing a culture that dates back to Ben Franklin is not any easy thing to do. Woodrow Wilson said, "If you want to make enemies, try to change something."

Okay, let's admit the truth to ourselves that we are not going to save everyone, and most of those fire fatalities that we see in our line of work were already a statistic before we even pulled out of the station.

When I look back at my two worst fatal fires, it still makes me wonder about the risk we took. The first was when I was a new firefighter just off of probation; that fire took the life of a mother and her son. It was midnight, and I was on the first arriving ambulance when we pulled up to a single family, 1,400-square-foot ranch with fire blowing out of almost every window. It was a cold night with winds blowing about 30 mph. There were flames blowing out of the downwind bedroom windows, which the neighbors had already broken out with the hopes of saving the mother. When the engine pulled up, we pulled a line and entered the house through the front door and headed down the hall toward the bedroom. We reached the bedroom, turned the corner, and could see the mother burning on the floor. That's when our hoseline burned through. As we exited the house, the second due engine was advancing the backup line, and the fire was extinguished within 30 or 40 minutes. There were no firefighter injuries. The bodies of the mother and son, who we later discovered on the far side of the bedroom under a window, were removed. Today, I know that both of those lives were gone before we even left the station. We were lucky that none of us were hurt trying to reach two bodies that were already gone.

My second call was also a small, single-family home, about 1,400 square feet, on a slab, with an attached two-car garage, fully involved. I was the company officer on the second due engine. We advanced a second line in through the front door and turned right down the hall toward the bedrooms. The first line had gone left into the living room/kitchen area. I remember a lot of fire in the hallway and overhead, which caused the ceiling to fall in on us. It was very difficult to crawl due to all the wires and stuff that were part of the collapsing ceiling. As we made the first bedroom, we found the badly burned bodies of two teenage girls on the floor. As we entered the second bedroom on the left, we found two more bodies, those of young boys still in their bunk beds. It wasn't until later, during overhaul, that we found the third boy who had apparently tried to hide in the closet. An older teenage boy was also discovered in a back bedroom by the laundry room. A total of six children, one fire, and I am sure they were all dead before we got the call. Although no firefighter suffered any physical injuries, I am sure the results would have been the same if we had just done a more defensive attack on the fire. Making that aggressive interior attack did nothing to save those children and only put our firefighters at risk.

So, is it worth driving lights and siren, blowing red lights, and putting everyone on the road in danger because we think that saving a few seconds is going to save a life? If we are being unsafe on the road just to save a house, it goes back to the thought that a house is nothing more than a place to store all our stuff. My point is that we shouldn't be risking our lives to simply save stuff.

I believe we have to take our job seriously. I also agree that it is an acceptable risk to save a life, but only a life. It is not an acceptable risk if all we recover is a body.

Thanks, Dad, for inspiring me to follow in your footsteps and join the greatest profession ever!

Kelvin J. Cochran

A Fire Chief Focused on Our Vulnerability

Kelvin has risen through the ranks from firefighter in Shreveport in 1981 to fire chief in 1999. He then served as fire chief in Atlanta from 2008 to 2009, when President Obama appointed him as U.S. Fire Administrator in Washington, D.C. In 2010 the mayor of Atlanta asked him to "return to quarters" and return to the position of chief in Atlanta—an offer he couldn't refuse. I met Kelvin several years ago when he had an idea related to "predicting" firefighter injury and death. His concept was to create a tool that would help a chief and department determine their level of vulnerability—a self-assessment of the organization to see what areas were weak and what were potential "problems lying in wait." That project, known as the Fire Service Vulnerability Assessment Program (VAP), was envisioned by Kelvin and was developed under the leadership of the NFFF. The VAP was funded initially by a grant from the U.S. Fire Administration and then exclusively by Honeywell Life Safety/First Responders. It will be made available in 2014 to every fire department in the country—at no cost. It is widely accepted that the VAP will absolutely be a game changer in the business of firefighting.

Kelvin J. Cochran

Transformational Followers

Over the past 30 years, the fire service has placed tremendous effort and energy on the need to develop leaders. Competencies for professionalism have been standardized in National Fire Protection Association (NFPA) 1021 for fire officers and chief officers. The Center for Public Safety Excellence has established professional designations for company officers and chief officers that confirm the attainment of NFPA competencies and other credentials through training, education, and experience.

Today, there are many options for higher education degrees for fire service professional development geared toward preparing members of the fire service for advancement and leadership. My career is one of many in this generation of firefighters who have benefited tremendously from taking advantage of the evolution of fire service leadership fostered by the United States Fire Administration's National Fire Academy and the International Association of Fire Chiefs.

After 13 years of serving as fire chief in Shreveport, Louisiana; Atlanta, Georgia; and as U.S. fire administrator, the leadership theories I have embraced to form my identity as a leader are: situational leadership, authentic leadership, servant leadership, and transformational leadership. In an effort to avoid lecturing, allow me to cut straight to the chase with a brief explanation.

Situational leadership is how we lead. Authentic leadership is who we are as we lead. Servant leadership is why we lead. And transformational leadership is what we are striving for as we lead. In other words, no successful leader leads the same way under all circumstances. We change our style based on the *situation*. Leaders who make a difference in their organizations are the right fit. We lead out of who we already are—*authentic*. We do not have to change who we are to be successful. Our motive for leading is to enhance our capacity to *serve* more and to impact the lives of more people. And our ultimate intent is to *transform* an organization to exceed the expectations of its stakeholders.

One the greatest expectations followers have of leaders is that leaders are to make things better than they were, no matter how successful their predecessor may have been. No one gets promoted to preserve the status quo. As such, followers expect transformational leaders. The primary focus of transformational leaders includes transforming culture, improving organizational behavior, enhancing performance management outcomes, accomplishing the mission, and pursuing the vision. Transformational leaders also focus on tangible change initiatives such as improving services to the community, enhancing member salaries and benefits, maintaining and adding fire stations, maintaining and replacing fleet, and enhancing professional development, just to name a few. Everyone wants a transformational leader.

Though there are no credible arguments against transformational leadership for the fire service—everybody expects and desires one—they are not always well received, and they are not always successful. Transformational leaders are often met with resistance. I believe one of the greatest contributing factors of why some transformational fire service leaders fail, and in some cases are deliberately derailed, is the failure of the fire service to focus on *transformational followership*.

Effective leaders are only as effective as the commitment and competence of their followers. Effective leadership requires effective followership. Transformational leadership requires transformational followership. Committed leaders initiate transformation, but it takes committed followers to build and sustain transformation momentum at the battalion and company level.

To shape the future of the fire service, there has to be a greater emphasis on the development of followers. Followership is foundational to leadership. John Maxwell says, "If you're out in front and no one is behind you, you are not leading; you're just going for a walk." Much emphasis has been placed on leadership—and rightfully so. But not enough emphasis has been placed on followership. To be effective in getting things accomplished we have taught, as previously stated, that leadership is situational. This approach is not always productive, because in today's fire service culture, followership is conditional.

We have evolved into a culture where followers get to choose, based on self-centered conditions, what they will do and what they will not do; what they will commit to and what they will not commit to. Worst of all, there are not consequences when their lack of contribution and commitment results in poor performance and inappropriate behavior, because we lack leaders who have the competence or courage to hold them accountable. Conditional followers resist transformation. They are against people, processes, and activities that challenge them to serve at a higher level and are against anyone who asks them to give more than the minimum requirements. Common traits and behaviors that distinguish *conditional followers* are:

- Conditional engagement with coworkers
- Conditional involvement in departmental activities

- Conditional enforcement of policies and procedures
- Conditional sharing of information
- Conditional compliance to directives
- Conditional pursuer of departmental vision
- Conditional commitment to departmental doctrine

The conditions assessed by this type of follower are quite simple and basically boils down to two questions, "What's in it for me?" or "How is it going to affect me?" When a department has too many members who are conditional followers, they generate a culture of indifference that stifles or kills meaningful transformation.

Transformational followers are a must for transformation to occur in a fire department. Members who expect transformational leadership without a commitment of being a transformational follower are hypocrites and are not serving for the right reasons. Their motives are not aligned with the values of what we stand for in the fire service. Transformational followers are committed to follow under all conditions. Rather than a self-centered philosophy, they have an others-centered philosophy. They have a what's-in-it-for-us mentality, not a what's-in-it-for-me mentality. Transformational followers:

- Interact with all of their coworkers
- Support all departmental activities and events
- Enforce all policies and procedures
- Share all information received with others
- Support departmental directives
- Actively pursue departmental vision
- Openly support departmental doctrine

As fire service leaders, we have got to get back to the basics of service, sacrifice, and followership. To follow means to come, or go after; to chase or pursue; to accompany; to go along the course of, to comply with or obey; to engage in; to ensure result and to pay attention to. These are the principles we have lived by over the history of our proud profession that have begun to diminish in recent years. Because most leaders have embraced the value of inclusion and participation in planning and decision making, some followers have falsely interpreted these opportunities as a right and not as a privilege, thinking they are entitled to refuse participation and support at their discretion. They withdraw their commitment to follow and yet maintain their expectation for the leader, particularly the fire chief, to be transformational with regard to preserving good tradition, changing bad culture and improving facilities, fleet, and employee benefits.

Our focus on professional development for leaders has served our profession well. However, we have not maximized the full benefit of leadership development due to the inadvertent lack of focus on followership development. Transformational leaders require transformational followers who not only have the knowledge, skills, and abilities to do their jobs, but also followers who have the confidence, commitment, motivation, and courage to initiate and sustain transformation momentum at their level within the organization.

The benefits of transformational leadership do not occur overnight, but result in substantial organizational rewards closely related to leader-follower satisfaction. How will we know when we have established a culture of transformational followership in our departments? We will know when the members collectively understand that the goals of the organization are of greater priority than their personal goals. We will know that we have arrived when the members as a whole are committed to obedience to departmental controls and culture, when it becomes unpopular to be a conditional follower. We will know we have an organization of transformational followers when by and large the members have a sincere desire to see each succeed.

In the future, when the history of the fire service is revised to capture the current generations' success stories, it will highlight the leaders of our present day who made a difference not because they achieved the highest rank or were the presidents and officers of our professional associations, but because they were servants—faithful *transformational followers*.

Allyson Coglianese and Eileen Coglianese

A Mom and Her Daughter: Literally Never Forgetting

In the very personal words of Allyson: "My father, Edmond P. Coglianese, was a lieutenant with the Chicago Fire Department. He gave his life, the ultimate sacrifice, doing what he loved and serving the citizens of Chicago on January 26th, 1986. He saw to the safe exit of several residents during an arson related fire at a transient hotel before returning again to search for more victims and succumbing to smoke inhalation and burns. I was 12 and my brother Matthew was 9. The difficult part of mourning my father's death was that I wanted and somehow needed everyone around me, everyone in Chicago, to be in mourning as well. But that was not to be. The Chicago Bears won the Super Bowl later that day, and now everyone's attention and priorities turned to celebrating the big win. Sharing this 'anniversary' of my hero's life and sacrifice with that of many Chicagoans' 'heroes,' a professional football team's championship that will never fade from the media's memory. Every year, every Super Bowl, thousands in and around Chicago reminisce and celebrate the 1985 Chicago Bear's victory. It is on the news, in the papers, it is everywhere. But only a few dozen or so remember and celebrate the life of a true hero."

Eileen serves with me as a member of the board of directors of the National Fallen Firefighters Foundation, where we have gotten to know each other. In 1991, she became one of the founding board members of the CFD's Gold Badge Society, a support group and advocate for families of duty death firefighters and paramedics of the CFD. In 2014, she will begin her 15th year as president of the Gold Badge Society. Her daughter Allyson serves as vice president. Eileen spearheaded the creation of the CFD Fallen Firefighter/Paramedic Memorial Park on Chicago's lakefront. Since the inception of the Everyone Goes Home initiatives, she addresses each firefighter candidate class to encourage compliance. She volunteers at the IAFF Fallen Firefighter Memorial as well as the National Fallen Firefighters Memorial weekends.

Lt. Edmond P. Coglianese
Chicago Fire Department, Engine 98
Last Call: January 26, 1986

Edmond Coglianese with his daughter, Allyson, during his training at the Chicago Fire Academy

My daddy spent the majority of his short 12-year career at Engine 98, and boy did he love that firehouse. He loved everything about the job, from helping others in their time of need to belonging to the "brotherhood" of being a firefighter. On a cold winter's morning, January 26, 1986, my father gave the ultimate sacrifice, his life, while fighting a 3-11 arson at a transient hotel. Lt. Ed Coglianese left behind his wife, Eileen; his 9-year-old son, Matthew; and me, his 12-year-old daughter.

Allyson Coglianese

I am in no position to tell anyone in the fire service how to do your job, as I am not qualified. But I do know what it is like to be given no choice, but to learn to accept the fact that my firefighter, my hero, my father would never come home from this job.

My dad's death was preventable.

Nearly every single line-of-duty injury and death is preventable. That is not to say that the job is not hazardous, because it is and yes, you have to be willing to put your life on the line. Just make certain that it is time to do it. The fire service does not have to continue with the mindset that it is all right to expect and accept that there could and will be injury and death in this profession. If tradition is killing, then tradition needs to change.

Now from time to time I wonder what really happened at the fire scene that fateful day. What was it like in that hotel? Did the conditions in the room my father was found in become so overwhelming that even the best equipment could fail to protect him? Did he make a mistake? Did someone else? And if so, what were they and why? Were shortcuts taken? I want to know what happened.

I do know that what happened in the following weeks was that the design of the mask he was wearing was altered. Department and federal investigations were conducted, of course; however, many of the reports filed did not corroborate. I believe that many wanted to make sure the situation looked good and that my family would be taken care of. But I also believe there were those who wanted to save their own hide. There was so much confusion around how the fire was started and the events that followed while fighting the fire that it is not entirely clear just what happened to my dad in that hotel room. What is known for certain is that it is extremely rare, if ever, that anyone dies when everything is done right.

The fire service does learn from the roughly 100 deaths that occur each year. There are always new health and safety standards being put into place. But all too often, they are not adhered to. Equipment has certainly improved over the years, but all too often it is not utilized properly. What we need is an attitude change and a change in behavior, and it starts with you. Through the National Fallen Firefighters Foundation, the *Everyone Goes Home* Life Safety Initiative program has made a difference in the fire service nationwide. However, one fatality is still too many.

Now to my knowledge, any mistake or mistakes made in relation to my father's death have been corrected by the Chicago Fire Department and that the circumstances of his death have been used in the implementation of and teaching of Rapid Intervention Team training. And I know that my dad would have wanted it this way. There will always be lessons to be learned and shared. There is always room for improvement and to continue your education. Accountability and responsibility, whether personal or as a team, go a long way.

How We Honor Our Fallen

In the 27 years since my husband, Lt. Edmond Coglianese, died in the line of duty, nearly 23 years involved with the Chicago Fire Department (CFD) Gold Badge Society (volunteer support group for LODD families), 15 years volunteering for the National Fallen Firefighters Foundation (NFFF) and the International Association of Fire Fighters (IAFF) Memorial, I have seen, heard, and learned a lot. When I am called upon to share my story, I generally focus on what I have learned from other families to be either helpful or hurtful to them as well, a commonality. I hold their stories forever in my heart. For this venue I have chosen to focus on just one aspect of dealing with families—honoring the family as well as their beloved fallen.

Lt. Edmond Coglianese died in the line of duty in Chicago in 1986.

The journey, of course, begins with a compassionate yet honest and timely notification. Not always, but generally, there is time spent in the emergency room. It is important to allow the family to spend as much time as needed with their loved one, time to notify other family members and those close to the deceased. Although necessary tasks, this is not the time to focus on how quickly the deceased is moved to the morgue and the name released to the press.

When offering to hold a LODD funeral to the family, try to make sure that they understand that it is an offer; they should be allowed to choose what aspects of the funeral they want or do not want included. It is important to help the family understand the meaning of the varied rituals that are generally included. Not only should the family be allowed to include personal wishes for the service, but should be encouraged to do so.

It does not end with the funeral. Be mindful of any tributes that you create in the firehouse. There may come a time when it becomes too difficult for those who need to work there to work in that environment, and the time to remove anything already created will never come.

As for any traditional memorial ceremonies, either local or statewide, again, please keep the families informed and give them as much advance notice as possible. Allow them to include extended family members and close friends. Many of these ceremonies are oriented toward the fire service. Those who serve in the fire service may appreciate hearing and even viewing the details of the incident, but it is almost always too difficult for the family to deal with this in a public place. Although the family may feel honored to have politicians attend, make every effort to keep the focus on the fallen who are being honored.

I have also seen some traditional annual memorials when not only those who have died in the line of duty are being honored, but other acts of courage and valor by those in the fire service are being recognized. This is unfair to all involved. For those who are being honored and present to accept that honor, it is a time for celebration, which they should be allowed to enjoy, not stifled by the necessary somberness of honoring the fallen. For families of the fallen, although most assuredly they would rejoice in knowing of such success stories, it is neither the time nor place.

Please feel assured that families know, understand, and are extremely grateful that you recognize the sacrifice and are dedicated to honoring them.

JOHN "SKIP" COLEMAN

The Street-Smart Firefighting Chief: From Blood, Sweat, and Tears

Skip is retired as assistant chief from the Toledo, OH, Department of Fire and Rescue. He is the author of numerous books and the 2011 recipient of the FDIC Tom Brennan Lifetime Achievement Award. I got to know Skip several years ago when we taught the original FDIC fire officers boot camp, along with FDNY Deputy Chief Jimmy Murtagh. Passion? Wow. He gets so passionate about fire operations, tactics, and being "smart" on the fireground that his passion oozes from his pores. Skip loves being on the fireground. Like me, he was born with a genetic defect that creates problems for us when we're around those who don't support what we do and how we do it. Read it in his books and hear him speak—Skip's passion for firefighting is contagious.

John "Skip" Coleman

My Advice to the Ranks

I am honored that Chief Goldfeder asked me to give my two cents worth within this endeavor. After much self-reflection, I have decided to break this essay up into three pieces of advice for the three levels of rank in the fire service.

After looking back over 32 years in the fire service, I have mixed feelings. Not about the job in any way. To me, it's the best job in the world. My mixed feelings concern me and my performance. In hindsight, I wish I had some do-overs in my career. A few at fires, but I am proud of the fact that the worst injuries firefighters received at my fires were a broken wrist (when a firefighter fell while descending a ladder from a roof), a good second-degree burn on a neck (I let a few good firefighters stay in a little too long at a vacant house fire), and a fractured bone in a foot (when a firefighter was on an aerial—not flowing water—when the driver accidentally moved the wrong lever the wrong way the firefighter's foot got caught between the rungs).

My do-overs would more involve dealing with other firefighters, especially after I made chief. I know that at one time in my career I was probably the most hated person in my department. I was operations deputy and doing what operations deputies do—trying to control all the personnel that ride the apparatus and respond to fires, EMS runs, and other emergencies. Apparently, I never got my copy of "The Operations Deputy Field Guide" and ran the personnel side of the job by the seat of my pants. I did the best I could. I made some errors and did some things I would now do differently, but I did the best I could. My one redeeming factor was that on several occasions, firefighters or officers I was dealing with would pull me aside and tell me, "You may be a jerk, but I always feel a little better when you pull up at a fire."

TO FIREFIGHTERS, PARTICULARLY RECRUITS AND NEWBIES

My advice: *Don't be worthless.* I can remember being a very young firefighter and listening to the conversations in the kitchen at the station. Where I came from, the day started at 0700 hours. After the rigs were checked and the gear was stowed, we would gather in the kitchen to catch up on what happened last tour and over the last two days off. Conversations centered on family, food, recent fires, and, of course, rumors. Much talk centered on other firefighters. Two terms were usually used to describe a firefighter. One was the good term. It rolled off the tongue and into the ear canal as "Oh yeah! Smith out of 14s—he's a good firefighter." The other term was the bad term. Again, it rolled off the tongue but pierced the ears as "Are you kidding me! He's worthless!"

How do you not be worthless? Be the first one up and the last one to sit. If you're loading hose, be the one up in the hosebed. If you're done eating dinner, be the first to start the dishes. If you're at a fire, do what you're told and then do some more. That's why they give us two days off, so we can rest.

TO COMPANY OFFICERS, PARTICULARLY NEW COMPANY OFFICERS

My advice: *Get off to a good new start.* Hopefully you have watched and listened as a firefighter. When my dad found out I was getting promoted to lieutenant, he told me, "Watch what officers do. The things that you like, do those! The things officers do that you don't like, don't do those!"

You know what your strengths and weaknesses are. Build upon your strengths, and fix your weaknesses. You have one chance to get it right, and this is the time. Even if you were as described earlier (worthless) as a firefighter, you don't have to be so as an officer.

TO CHIEF OFFICERS

My advice: *Lead by example* and *be consistent*. First, lead by example: If you make them do "it," then you better by God do "it," too! Bunker gear, SCBA, seat belt, neck tie in the summer, and no hoodies in the winter. You know the rules and are sworn to follow them (did you ever actually read your oath of office?). Nothing will tarnish your reputation as a chief more than by enacting the "don't do as I do, do as I say" attitude. Nothing will bring on the "oh!—he's one of those" behind your back.

Second, be consistent. You can't pick and choose the rules you intend to enforce and follow. If there's a dumb rule, then work to get it changed or eliminated. You can't write up one firefighter for using inappropriate language in the station and then the next day listen to another firefighter say the same thing and when asked about it say "oh, that's just Joe being Joe."

Ronny J. Coleman

Defining "Making a Difference"

I have followed and read Chief Ronny Coleman since I was a young firefighter. There are few who have the energy and passion that he has for being a firefighter. I first met Ronny when he was working on a smoke detector project—after some children were killed. We have remained friends, and he is absolutely one of the smartest people I have ever met—often seeing things that the rest of us simply don't. One of Ronny's programs that really stands out, created well before its time, was his "Making a Difference" fire officer training program. A bit "Ben Franklin-ish" with his homespun wisdom, Ronny has so many great ideas on how we can improve ourselves—and those we serve—and has been giving back to the job for decades. An IAFC past president, former California state fire marshal, active in numerous committees and boards, Ronny is the author of more than 19 textbooks for the fire community and is a veteran chief fire officer who has been making a difference for so many of us . . . for so many years.

Ronny J. Coleman

The Bondi Story

In my previous contribution to Chief Goldfeders's work, I told the story of two fires that shaped my lifetime philosophy of support for built-in fire protection. After I left Costa Mesa Fire and became fire chief in San Clemente in 1971, I had additional reasons given to me for why I needed to pursue a goal of making homes safer from fire.

Upon taking command of the San Clemente Fire Department, I was confronted with a citywide general plan that indicated our city was going to multiply four times in the next two decades and almost all of it was going to be residential occupancies. It provided an excellent laboratory for me to put into practice what I believed in in as a practicing fire officer. Little did I know, however, how many other people would participate in encouraging my activity to become even more focused on the goal of built-in fire protection.

The first name I would evoke was that of Fire Chief Richard Bosted. He was among the very earliest advocates of smoke detectors. When I first met him, he was fire chief of Brea, California. Although he supported my efforts at researching residential sprinklers in the early days, his real focus was on smoke detection. Richard never lost an opportunity to lecture me on the value of early-warning devices. At that time, he was actively involved in an organization called We Tip. He was the vice president and director of the organization. Whereas its focus was on arson reduction, Richard managed to leverage the program to discuss smoke detectors. Among his proudest moments was convincing a movie star named Iron Eyes Cody to be a spokesperson for smoke detection.

His real contribution to my experience, however, was introducing me to Ray Jewell. Ray was a movie producer in Hollywood. He produced public education films along with a line of other documentaries. Richard got me in touch with Ray because he had just finished a film called "The Bondi Story." Unfortunately, it was a story with a common theme. It involved the death of the Bondi family children in a fire in which no smoke detection had been provided. As a direct result of the introduction, Ray Jewell and I became pretty good friends. I used his Bondi story extensively in my public education program, and as we moved forward with our research on residential sprinklers, Ray expressed interest in doing a second film that would emphasize how residential sprinklers could also contribute to life safety.

Talking with the Bondi family was sometimes difficult. It was hard to carry on a conversation without causing someone's eyes to well up with emotion. Nonetheless, we proceeded with the development of a script and set a production schedule into place.

Fortunately at the same time, the City of Burbank was getting ready to burn some buildings. Working closely with the Burbank Fire Department, Ray Jewell and I went to work setting up a controlled burn in which we were going to burn two rooms side by side to demonstrate the effectiveness of fire sprinklers in reducing the danger to the occupants. At about that same time, I was also in conversation with Chief Don Manning of the Los Angeles Fire Department who was doing research on the speed with which sprinkler heads responded. This was before they had developed the concept of the quick response sprinkler (QRS).

The team that worked to put this film together spent countless hours getting ready for the big day. Writer Elizabeth Grumette and her writer/director/camera operator husband, Steve Grumette, Ray Jewell, and I became very close to the Bondi family.

Little did I also know that that day would introduce me to another lifelong friend. He also happens to be the editor of this book.

One of my closest friends at that time was Warren Isman. He had introduced me via phone with a young fire officer by the name of Billy Goldfeder. I do not recall all of the details of how we actually ended up with him arriving at my office at the same time we were getting ready to do production. This day in 1979 proved to be another of those days that I recall as being pivotal in strengthening my advocacy in sprinklers.

In retrospect, I consider that day to be another turning point in my career because it demonstrated to me how powerful messages are when they are packaged appropriately and distributed effectively. On the other hand, it was a watershed date for me because we had linked up with an individual who would someday become one of the most influential communicators in the American Fire Service: Billy G.

Paul Combs

A Picture Is Worth a Thousand Lessons: Helping Us Become Better Firefighters

How about that picture on the cover of this book? They say a true artist is able create a masterpiece when given an impossible subject to work with. That's Paul Combs, miracle worker. A long time friend, Paul "does us a favor" and looks closely at us (for us!) as firefighters, and as human beings, which is important—and never easy. Sometimes we need a little help. The help came to the world's fire service in the form of this incredibly talented firefighting artist Paul Combs. Paul is one of those people who say (through his amazing artwork) what we are thinking—only no one wants to say it. And just when we are getting comfortable, Paul's art grabs us by the head and jolts us back into reality... and into even more critical thinking. Paul is a fire lieutenant and veteran firefighter for the City of Bryan (OH) FD, a firefighter II, NREMT-B, fire officer I, hazmat technician, and Ohio fire instructor. He's an artist who does what you do. *And how about this:* he can also write, and here are some words from Paul.

There are different leadership styles, but leaders who bully employees are never successful.

No Place for Bullies

I hate bullies—always have, always will! Unfortunately, this is a leadership trait of far too many company and chief officers who rule their firefighters with a domineering management style. They focus on the negative and all too often take positive and encouraging actions for granted. These officers are impossible to please; they cause high employee turnover; and crew and individual potential is never reached. Domineering leaders hide their lack of knowledge and social skills behind aggression and intimidation. As with most bullies, insecurity plays a huge role in this type of firefighter management, as they tend to make decisions based on their personal needs and ambitions rather than the good of the company. Regardless of their reasons, these people are impossible to work with and do much more harm than good.

Great leaders motivate and empower their firefighters to excel. People enjoy working for fair leaders, and they have the highest productivity and retention rates of all the management styles. Fair leaders can be tough and demanding (there's nothing wrong with that), but gain respect due to their ability to justify their decisions with sound strategy, knowledge, experience, and most of all, a show of respect to others. Great leaders lead from the front of the pack, they guide and strengthen their people with a strong and steady hand that is unshakable during turbulent times.

So, what are traits of great leaders?

Great leaders are effective communicators. Your subordinates must feel like they can come to you for direction and guidance, and you must be willing and eager to listen. You may not have all the answers to their issues, but just being there will go far in gaining their trust. Remember, a firehouse is a close social group, and others will notice when you simply take time to listen.

Great leaders value every member of the team; in turn, your firefighters know they are valued, respected, and important. Treat every firefighter as you would like to be treated—and never forget that you were once a probationary firefighter, too.

Great leaders take ultimate ownership and accountability for when things go wrong—and things *will* go wrong from time to time. Respected leaders give ultimate ownership and praise to their crew when things go right. Never, ever play the blame game—criticism comes to you, praise goes to them.

Great leaders are role models for their subordinates. Everything they do is a reflection of how firefighters are expected to work and behave. From your appearance, actions, work ethic, and integrity—act how you want your employees to act. They are always watching you—give them an example to follow.

Great leaders understand they can never know everything, and there is always something new to learn. A great leader looks to their crew and superiors both within and outside of the firehouse for inspiration and knowledge, and shares that passion for learning with their firefighters. When your firefighters see you studying and learning new ways to better yourself and them, they will aspire to follow your lead. Show them praise for their willingness to learn, and encourage the others to follow the same path.

Lastly, great leaders never ask their firefighters to do something they wouldn't do themselves

No one wants to work under the watchful eye of a tyrant, and never underestimate the effects of poor morale within your department. Any military commander worth their salt will tell you that poor morale is like a poison, and once it starts to spread it can kill a healthy, well-trained group of soldiers—firefighting is no different.

Don't be the bully!

Dennis Compton

Cool, Calm Local and National Leadership

Chief Denny Compton is one of those cool, calm people. Don't let him fool you—he is as gung-ho about this stuff as much as anyone. The difference is that he has mastered staying cool and calm, which allows him to get things done. It also allows him to get others to get things done in his many leadership roles. His "come on—let's go!" attitude works quickly, especially when bringing our very diverse fire service together to accomplish some common needs. During more than 43 years in the fire service, Denny has served as fire chief in Mesa, AZ, assistant fire chief in Phoenix, AZ, chairman of the executive board of the International Fire Service Training Association (IFSTA), chairman of the Congressional Fire Services Institute (CFSI) National Advisory Committee, and is currently chairman of the National Fallen Firefighters Foundation board of directors. Author of many leadership books that have become "standards," Denny is the recipient of many awards and accolades. His attitude shines through on whatever he is involved with—including the people, who benefit most from his dedication.

Dennis Compton

Being a Firefighter Is a Privilege to Be Cherished

I wasn't one of those recruits who had planned for most of my life to become a firefighter. In fact, I had just been discharged from the U.S. Army and was trying to decide what to do with my life. I ran into a friend from high school who had been a Phoenix, Arizona, firefighter for about six months, and he suggested that I take the upcoming test—so I did. The next thing I knew, it was less than a year later and I was in the Phoenix Fire Department Training Academy learning to be a firefighter. Just as my high school friend had predicted, it was something I fell in love with immediately, and being a firefighter is one of the most cherished blessings of my entire life.

It didn't take long to learn a couple of timeless things about the fire department. After just a short time, the more senior firefighters and officers helped me see that there were (for the most part) two dominant types of members of the department:

- Givers: Those who liked the job, took great pride in being a firefighter, did whatever it took to get things done, and willingly did whatever was possible to solve the problems of the people who called the fire department.

- Takers: Those who didn't care that much about the fire department, or even other firefighters, took everything they could from the job, complained regularly about having to do the work and deal with the public, stayed away from the department as much as they could, and gave nothing of themselves to the fire department that didn't result in some personal reward.

I came to realize that there are varying degrees of givers and takers if you put them on separate giver/taker scales of 1 to 10. I also saw that if someone was revealed as a dominant taker while they were still on probation, they were usually eliminated from the job at that point.

Careers are long, and during a career a lot of things happen to and around firefighters. Most of the firefighters I have known deal with the ups and downs of the job quite effectively throughout their career. They positively identify with being a firefighter, and their pride is evident. Their bodies grow older, some promote and some don't, but they tend to become wiser and stay trained and fit so they can perform as effectively and safely as possible on the job. But sadly, that is not the case with some of the firefighters I've come across.

I have come to believe that nothing affects the quality of a person's fire service career more than individually coming to grips with this issue of givers and takers. Serving as a firefighter and being part of the fire service is a privilege—it's not a right or an entitlement—and the responsibilities we have to each other and the people we serve should never be taken lightly nor for granted.

You see, no matter what rank we achieve during our time in the fire service, we get to decide this giver/taker thing solely for ourselves. The decision you make regarding this issue will significantly impact your self-image; your sense of self-worth; your attitude; your performance; your ability to change and adjust to things that occur; your ability to cope with the stresses that accompany the job—and, for that matter, life in general; the level of respect others have for you; and even your ability to stay safe and survive. Bottom line, when we decide to go on the fire department, we commit most of our adult work lives to fulfilling that mission. It is our responsibility to stay positive, productive, and healthy for as long as we stick around in the fire service—from the first day to the very last day. Nothing affects the quality of your career more than whether you brand yourself as a giver or a taker in the eyes of others in the fire service. As firefighters, we need to come to understand that early on—never forget it—and pass it along to others.

David "Chip" Comstock, Jr.

Not Just Another Lawyer

I've gotten to know Chip over the years as a friend, as both a really gung-ho fire chief and a solid attorney who has defended firefighters against the forces of evil—sometimes even against city hall dwellers. My hero. He's been a firefighter for 30 years and is the chief of the Western Reserve Joint Fire District in Poland, Ohio. Chip is also an attorney in the firm of Comstock, Springer & Wilson Co., L.P.A., in Youngstown, Ohio. His practice focuses on fire department operations and general litigation, including governmental liability and insurance fraud/arson cases. Back in 2010, I wrote about an explosion his department experienced that injured some members. Chip shared this close call so we could all learn, remind, and remember . . . and there is much more we can learn from him.

The seven District firefighters involved in the residential gas explosion: (left to right) Terry Ferrick, J.R. Warren, Steve Dubic, Al Rivalsky, Troy Stewart, Scott O'Hara, John Walsh

Caring

I have now spent more than 25 years as a fire service lawyer. In that role, I have had the opportunity to work with some great chiefs, but I have also had to represent too many fire chiefs in disciplinary actions or in conflicts with their superiors, subordinates, or the public. I have also represented employers on numerous occasions when a less-than-stellar officer had to be removed.

When I first thought of providing advice in this book, I was going to write from a barrister's prospective. However, the more I thought about my legal involvement in employment conflicts, the more I realized that what I had learned in more than 30 years in the fire service from the officers that I served under (both good and bad) would provide a better foundation to give advice to keep fire officers out of legal trouble. However, my advice here is directed toward the newly appointed company officer, because the behaviors they learn and emulate at that level stay with them as they are promoted through the officer ranks. My chapter is about caring. If the proper caring attitude and behavior are adapted early, the future chief will never need my legal services.

First, fire officers must care about themselves. Fire officers care about themselves by making every effort to further their education, training, and experiences so as to improve their knowledge and competence. Firefighters, whether volunteer or career, can be professionals only by being experts in their field. An expert is defined by the federal rules of evidence as an individual who has specialized education, training, or experience in a particular field. All firefighters should be experts, and they should understand that the information we rely on in the management and operation of a fire department changes on a regular basis.

Lawsuits against fire officers and fire departments are becoming more frequent. The public's expectations relating to the fire department are as high as ever. That is not necessarily a bad thing. However, when a firefighter is seriously injured or killed in the line of duty, the family will look for compensation, retribution, or both. Likewise, members of the public who lose loved ones or property will not be forgiving. In those instances, fire officers are required to take the witness stand after the lawsuit is filed. Lawyers who are knowledgeable in fire operations soon separate the expert firefighters from the novices. Previously, fire officers who were ignorant of national standards have been charged criminally as a result of fire ground deaths. In New York State, a fire recruit was killed in a live burn training exercise. The officer in charge of the exercise was criminally charged and eventually convicted as a result of the fire training death. The officer was found to have violated NFPA 1403 (the national standard for conducting live burn training exercises in acquired structures). Other fire officers have also been charged criminally or sued civilly as a result of their fire ground conduct. The choice for a fire officer is clear—you can remain a lifelong student of the fire service and serve as an expert, or you can sit as a novice in the courtroom at the defendant's table.

Fire officers also care about themselves by being physically fit. Being a firefighter requires a high level of aerobic fitness, muscular endurance, and strength. Yet, firefighters frequently lack the level of physical fitness deemed necessary to handle rigorous occupational stressors. As a result, approximately one-half of annual line-of-duty deaths are attributable to cardiac-related problems. Both the International Association of Fire Chiefs and the International Association of Firefighters have recognized the need to promote health and fitness programs within the fire service. Nonetheless, there is a significant resistance to this movement, and many unions or fire associations have opposed any effort to make physical fitness standards mandatory. However, fire officers do not need to follow mandatory standards to be physically fit. They should be willing to set the example within the firehouse by exercising and by encouraging crew members to do likewise.

I have always tried to encourage others to participate in athletic activities. I like to run (more like a slow jog these days) and bike. I often ask firefighters to join me, and during the last few years, department members have participated in several 5K runs together, including one that involves obstacles and lots of mud. One year, we even played the local police departments in a charity tackle football game (we won 44-0, by the way). I trained for each of the events. Not only did my own fitness improve, but so did the other firefighters. It also improved fire department morale and teamwork. The board of trustees—our governing body—recognized this as well, and now pays for each firefighter to participate in various athletic events.

Fire officers who care about themselves also recognize that they accomplish more with their crew by remaining mentally positive. A positive outlook with respect to the department makes life much happier. I have visited many fire stations where the officers had positive attitudes, and as a result, the firefighters enjoyed working in that environment. Conversely, I have worked as an attorney in many fire departments where a company officer was insufferable and ultimately was removed. Again, I don't have to look far for examples of poor leadership. In one case, a company officer screamed all the time. I believe that screaming shows a lack of confidence and control. Too often, an officer's screams are perceived by employees in the department negatively, but unfortunately, that negativity finds a way to invade the entire station. In a volunteer department, just watch how attendance falls proportionally when screaming increases.

Attitude affects those around the officer, and ultimately impacts the health of not only the individuals but the organization as a whole. With respect to attitude, fire officers must remain humble, happy, and enthusiastic. I am a graduate of the Ohio State University. In former coach Jim Tressel's book, *The Winner's Manual for the Game of Life*, he set forth nine principles relating to having a proper attitude. Although written for athletes, the principles apply equally for fire officers and their crew:

1. More athletes fail through faulty mental attitudes than in any other way.
2. Attitudes are habits of thinking. You have it within your power to develop the habit of thinking thoughts that will result in a winning attitude.
3. The foundation for the proper attitude consists of developing the habit of thinking positive thoughts.
4. Tell yourself constantly that you can do something, and you will. Tell yourself you can't, and your subconscious mind will find a way for you not to do it.
5. A desire to win and a desire to prepare to win are important ingredients of a winning attitude.
6. Before you can scale the heights of athletic greatness, you must first learn to control yourself from within. Be your own master. Control your emotions.
7. An athlete with a good attitude is coachable. He welcomes criticism, constantly seeks to learn, and avoids criticizing his coach or teammates.
8. True success depends on teamwork, and the winning attitude puts the good of the team ahead of anything else.
9. Whether or not you create a winning attitude is entirely up to you - but nothing is more important to you on your road to the winner's circle.

Officers care about themselves by understanding that they do not have sufficient time to complete the tasks necessary to run the fire station. The tasks that need to be completed on a daily or weekly basis within a fire station appear to be endless. Conversely, an officer's time is finite. Fire officers who care about themselves delegate to others the responsibilities associated with running the crew or fire house. The officer feels less pressure to try to accomplish everything at once, and at the same time, provides opportunities and responsibilities to subordinates.

One of my mentors is Ed Chinowth, the chief who preceded me at the fire district. I joined the district, in part, because he believed in education, was progressive, and took an interest in everyone's well being. However, Chief Chinowth did not want to "burden" others with the many tasks of running a fire department. He knew that our volunteers had family and work commitments (the two things he always said came before the fire department). While I love Ed for all he has done for me, this is the one area where our management styles differ significantly. I delegate every chance I get. We have created 15 committees that address all facets of the fire department operations, including health and safety, EMS, training, retention and recruitment, stations, apparatus, equipment, fire prevention, fire operations, and technical rescue. Every member is expected to serve on at least one committee, and as a result, I am always amazed at how much gets accomplished. Our fire prevention committee—one of the most active—is constantly receiving requests from the schools for various programs. I could never have accomplished so much by myself and as a result, my stress level is much less.

More importantly than taking care of themselves, great officers take care of their crew. This task begins by getting to know your subordinates by building relationships. Fire officers must know the people they work with—what their interests and desires are both on and off the job. A great officer recognizes a subordinate's deficiencies, and lays out a plan to improve the firefighters' skills. It has often been said that "talent knows no rank" and an officer should be able to discover and cultivate the individuals' talents. All firefighters must first be taught the basic firefighting skills and must learn them well. Time must be taken to teach firefighters how to force entry, search, and ventilate structures. This takes company officer time and commitment, and depending on the student, is not always an easy task. But after these basis skills are mastered, additional interests can be developed and the firefighter's career advanced. UCLA basketball coach John Wooden wrote that "Each member of your team has the potential for personal greatness; a leader's job is to teach them how to do it." The most effective leaders are good teachers.

Individual development is further encouraged by delegating tasks and empowering the subordinate to make decisions and to accept responsibility on a regular basis. As I stated previously, our fire department has many committees, and everyone is expected to serve. Each committee is chaired by an officer or firefighter. These committees give many individuals the opportunity to take on responsibility and to demonstrate what they can do for the department and community. A company officer should not only be willing to accept additional responsibility to develop their leadership skills, but they should also be willing to encourage the crew members to take on additional responsibility and tasks as well—such as preparing a training drill, drafting a new procedure, or assuming responsibility for pre-planning a building.

When a company officer delegates, the results are not always successful. But many times they are. Officers who have faith in their subordinates take the opportunity and the risk that comes with delegation. A great officer does not accept credit where the positive outcome is a result of their subordinate's efforts, but instead gives credit where credit is due. A caring officer also expects that crewmembers make mistakes, and accept responsibility and blame when negative outcomes result. A great officer is respected when they protect the crew from unnecessary outside harm or criticism. This encouragement, delegation, and protection builds the trust necessary for the leader and crew to succeed as a team.

A great officer ignores trivial issues, but practices fair, firm, and consistent disciplinary actions. A great fire officer puts personalities aside in doing so. I previously mentioned an officer that had a propensity for screaming at firefighters. One thing about both career and volunteer firefighters—they don't want to make mistakes and often know when they have made one. When a mistake is made, often the firefighter only needs to be asked what he did wrong and what he would do differently next time. For more serious infractions, the proper course of action is to proceed slowly and deliberately, and to investigate all the facts so that a fair and informed decision can be reached and the appropriate action taken. All firefighters, even those guilty of serious offenses, want to be treated fairly.

One of my jobs as an attorney is to prosecute lawyers for ethical misconduct. In ethics disciplinary cases, a lawyer's law license, and livelihood, is often at risk. As a rule, lawyers can be a contentious lot. But I have had very few cases in which a lawyer became upset with me during a prosecution. That is simply because I always treated a lawyer as I would want to be treated. I use the same process when I am involved in a firefighter disciplinary case, whether as a lawyer or a chief officer. I never hide facts or information from a firefighter that I believe has committed misconduct. I always ask for an explanation, and I always remember that I am seeking the truth. Believe it or not, I sometimes find that I don't have all of the facts, and yes, occasionally I make a mistake. You must always keep in mind that although you may be dealing with one firefighter's disciplinary case, the entire crew or department will be watching how you treat that individual, and always assumes that you will treat everyone that same way.

A caring fire officer watches out for the *probie*. They recognize that the fire service is neither a college fraternity nor a Marine boot camp. I have occasionally witnessed company officers (or training instructors) who have apparently watched the movie *Full Metal Jacket* a few too many times, and have somehow been transformed into Gunnery Sgt. Hartman. These fire officers do everything but put a donut in the new recruit's mouth! I guess these officers never watched the entire movie, for if they had, they would have noticed that the marine recruit later shot and killed the sergeant.

I understand that firefighting is a stressful job, but treating new recruits poorly, whether by physical or mental abuse, is inexcusable. There are many ways in which to make recruits better firefighters. Recruits should be treated in a fair and objective manner and should be able to handle the stressors of the job through repetitive and realistic training. I also disagree with anyone who states that members are made to feel a member of the team by being hazed. Harassment or hazing does not constitute a team-building exercise.

In his book, *It's Your Ship*, Michael Abrashoff wrote:

> "In battle, our initial reactions can often be the difference between success and failure, life and death. We also need to apply successes and failures of others to our own situation and learn from them. If you prepare for the most challenging scenarios, chances are good that you will be much better prepared for the unforeseen."

Abrashoff also noted that if a portion of the crew was found to be weak when completing training, which was purposefully tougher than anything the ship would ever see in actual combat, those members received remedial training apart from the others, which saved time and strengthened their confidence and skills.

In today's world, fire officers complain too often about the difficulty with volunteer recruitment and retention. A volunteer's time is valuable, and they will invest the time where they receive the greatest rewards. Every firefighter, including the recruit, should feel like not only a member of the team, but a member of the family, dysfunctional as it may be at times.

In addition to taking care of themselves and their crew, a great officer takes care of the community. The officer spends sufficient time to build relationships with neighbors and community leaders. Too often, members of the fire service forget why we are here. They worry only about work, the pay that they receive, and the last directive by the chief that will somehow make their job harder. They forget that the public pays for their services, whether for wages or equipment. They complain that members of the public don't recognize the need for additional firefighters, newer equipment, new firehouses, or money for training. Yet, they refuse to go into their communities to win the minds and hearts of the residents they serve and protect with the understanding that will eventually result in additional votes to pass tax levies to add additional personnel or buy new equipment.

In our fire district, the firefighters association has sponsored a fundraising effort every December whereby it raises money for the purchase of food and gifts, which are then distributed anonymously to needy residents of the community. Extensive fundraising efforts are undertaken, and the firefighters work with local churches and schools to identify those in need. We do not receive thanks from those we help, but the community has become aware of the efforts undertaken by the members of the fire department and this good will has contributed to our ability to successfully pass tax levies.

A great officer recognizes the diversity of the community (culture, socio-economic, and racial). They will do whatever they can to build alliances. While acknowledging those groups, the caring officer will focus on the people we serve within the community. Marian Wright Edelman wrote in *The Measure of Our Success* that we must "help America remember that the fellowship of human beings is more important than the fellowship of race, class, and gender in a democratic society." The officer works toward unity of community that the firehouse serves. I am friends with a fire officer in a nearby urban city. She has organized a group of career African-American firefighters that work with a city elementary school to provide both school supplies and tutoring. The results within the school and the local community have been well received. The organization has also begun awarding scholarships to college for graduating intercity high school seniors. These firefighters go beyond the duties within the walls of the firehouse, and put their time and money back into the community where they serve as positive role models for children of both sexes and all races.

Members of the Fire District accepting the Ohio Volunteer Fire Department of the Year award in Columbus, Ohio: (left to right) Gio Melia, Bill O'Hara, Barb Stewart, David Comstock, Jr., Michelle Lumpp, Jim Stewart, Nancy Chinowth, Ed Chinowth, Diane Ingold, and Lee Ingold

Like with their crew, the caring officer is honest with the community, and while taking pride in the fire department's strengths and accomplishments, is honest with the community with respect to the crew's or department's shortcomings. Recently, I attended a presentation where an officer informed members of a city council about the difficulty with response times due to a lack of volunteers. He advised them candidly of the risks that were associated with fewer firefighters during daytime emergencies within the town. Laying our cards on the table can be difficult at times, and even embarrassing. However, remembering that we serve the community and that honesty should be one of our priorities, citizens have the right to know when a department can only fulfill a portion of its mission. When fire officers identify the deficiencies, they must then work with members of the crew, department, and the community to find solutions. Company officers must start from the beginning to identify deficiencies within the crew or department operations, to share those deficiencies with others, and to become part of the solution in bettering the department.

Finally, great and caring company officers understand that when they were promoted, it was not a coronation process. I view a promotion simply as the acceptance of additional responsibility for more firefighters' well being and safety. Initially, each firefighter is responsible for his or her own safety. As a company officer, you may be responsible for the safety of one to five other firefighters. As a chief officer, you may be responsible for 50 to 100 firefighters' safety, depending upon the size of an emergency. Ultimately, whether a firefighter returns home depends on the ability of the team to recognize hazards and to perform efficiently. A caring officer always remains humble and attempts to recognize and improve on their shortcomings. At the same time, they will also work to improve the capabilities of each member of the team, as well as the department as a whole. Improvement within the department requires working with all members of the community while never forgetting why the caring fire officer entered the fire service in the first place.

Glenn P. Corbett

Don't Just Do the Job, *Do the Job Well*

Glenn Corbett kind of looks like a professor but speaks like a firefighter. Educated writer, fire chief, and gentleman, Glenn has been properly described as a walking encyclopedia of the fire service. Glenn is an associate professor of fire science at John Jay College of Criminal Justice, but more importantly, Glenn is a tireless warrior. He speaks out on complex fire protection issues and translates them clearly for the press. His expertise and incredible attitude led the families of 9/11 to get their cries for help heard. Glenn testified before the 9/11 Commission and recently served on the Federal Advisory Committee of the National Construction Safety Team that investigated the World Trade Center attacks and created the codes and response procedures. Glenn has been an unsung hero of the fire service for decades, speaking up for those who can't or won't—and when he speaks, people listen.

Glenn Corbett

Know Your History!

The late fire service legend Francis L. Brannigan admonished firefighters with a singular call to action: "Know your buildings!" In essence, he highlighted the critical need for firefighters to understand how buildings are constructed and how they behave under fire attack. Most importantly, he wanted individual firefighters to study the buildings in their own response areas in order to understand them and identify their vulnerabilities, including their collapse potential.

Frank was a mentor to me and many others (you should have your own fire service mentor, a trusted, experienced person to guide you in your career). He was a visionary who passed on his knowledge and experiences to all who came in contact with him, either by attending his lectures or reading his classic text *Building Construction for the Fire Service*. We all owe him a debt of gratitude.

Building on this cornerstone of firefighter safety that Frank created, I'd also like to admonish you: Know your history!

Fire service history—and the traditions that have sprung from it—is replete with stories of tragedy and triumph. It's very important for every firefighter to know—and be proud of—the experiences of our past. It's what makes us unlike virtually every other group, save the military.

From a practical perspective, knowing our fireground history is essential to our collective safety and effectiveness. There's an old adage from the American philosopher George Santayana: "Those who do not learn from history are doomed to repeat it." This rings true, particularly when we look at fires which have claimed the lives of firefighters under circumstances very similar to those of past fires.

Although most of us are familiar with the hazards of modern lightweight construction including wood trusses with gang nail connectors and wood I-beams with paper thin webs, we can't forget about the older buildings we encounter. In some communities, buildings more than 100 years old are the norm rather than the exception. Given their age, these types of buildings have decades of fire experience.

Consider the Brockton, Massachusetts, theater fire and collapse of 1941. Although this fire occurred before many of us were born, it holds hard-earned lessons that are still relevant today. The fire began in the basement of the Strand Theater, ultimately spreading up 60 feet through the walls to the roof. The fire, consuming wood joists supporting the balcony ceiling below and heating the unprotected steel roof trusses above, collapsed onto firefighters operating underneath. Thirteen firefighters died as a result.

Is this a dusty fire of the past, one that could never happen again? Of course not! We are surrounded by old buildings—some identical to this one—all of which hold the potential of burning and collapsing just like they did over 100 years ago. Their fire history holds an enormous set of lessons for us, if we only look and find them. It's incredibly valuable taking a look at fires of the distant past as well as those that occurred very recently.

Don't confine your historical review to just fires in buildings. Look at our history with other kinds of fires and hazardous materials. Read and think about the 1988 explosion of ammonium nitrate and fuel oil that killed 6 firefighters in Kansas City, Missouri, and the recent West, Texas, ammonium nitrate explosion that killed 11 firefighters. Research the 1973 Kingman, Arizona, boiling liquid expanding vapor explosion (BLEVE) and the 1994 Colorado South Canyon fire. With the incredible power of the Internet, most of our fire history is now at your fingertips.

Take the initiative, and study the kinds of incidents that could happen to you and your department tomorrow or next week. Distill and collect the hard-earned lessons that were learned. Put them in a presentation format and make a drill out of them. Besides being a potential lifesaver, your efforts will be a tremendous legacy to those who made the supreme sacrifice at these incidents of the past. There is no other better memorial to their memories than to learn their lessons and pass them on to other firefighters.

Know your history!

Dave Daniels

Rethinking Our Fire Service

Dave and I got to know each other when he became involved with the International Association of Fire Chiefs Safety, Health, and Survival Section. At the time, he was the fire chief in Fulton County, Georgia. From the time we met through our continued friendship today, I looked at him as a chief who definitely wants us to "rethink" how we operate. Dave has taken many of the lessons he learned in the fire service safety movement into his current capacity as the executive director of safety in the office of Atlanta Mayor Kasim Reed. Atlanta is one of the first cities in the United States to establish a workplace safety position in the mayor's office, and they've chosen an experienced safety professional who also happens to be a former fire chief to lead the charge. Dave's thoughts these days are more toward the "human" aspect of the job, including human resource management, occupational safety, and organizational culture. Think that's not "sexy" stuff? Just like water, we can't operate without it, and Dave offers so much for us to think about.

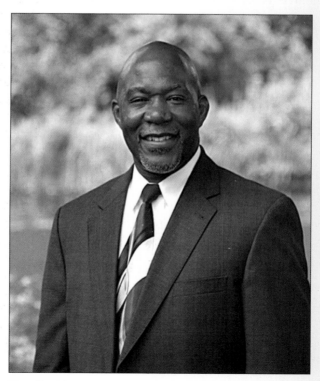

Dave Daniels

It Is Okay to Keep Studying

While the fire service has a long and storied history of courage under fire and exceptional deeds in unique circumstances, the service is not generally known as the most intellectual bunch. For those who have just taken offense at this statement and decided to go and read another story about a big fire, I have achieved my goal. Those who are intrigued by the word *geek* and a little bored with the "big fire" stories, allow me to share with you something I learned over a 30-year career in the fire service: it's okay to keep studying.

Often, young wannabes enter the recruit training process uncertain about exactly *what* they will get out of it and in some cases wonder *if* they will get out of it. The recruit training process is in many cases the point when firefighters are at their best. They are physically in the best shape of their lives. They are emotionally the most inspired and the hungriest for knowledge as they will likely ever be. So what happens over the course of a long career? Many simply stop learning, and that happens because they stop studying. In recruit school there is an expectation that every day is going to be an evaluation or a test of what you learned the day or the week before. This often drives a new firefighter to study their craft in a deeper and more reflective manner in response to the constant scrutiny of the recruit training process. Those who complete the process do so not because they are inherently gifted, but because of the hard work that they put into studying the various subjects that they are faced with including basic fire behavior, hose, ladder, and tool manipulation and the intricacies of hose and ladder evolutions.

The second thing I learned is it's okay to keep studying. After you leave the recruit training academy and receive an assignment to a fire company or ambulance, depending on the system, making it through the probationary or the initial enculturation process requires that you continue to prove yourself to your crew. Proving yourself in recruit school is about knowing the facts and the figures. Proving yourself in the company is about knowing your colleagues, the facts, and the figures. Believe it or not, you will need to study the people that you work with. Ask yourself the question, "Do I want to be what this group is?" "Do they have the same drive that I have to be the best I can be?" "Do they have the level of determination that I have?" To determine all this requires study. You may think you know the group, but don't jump to conclusions.

The professional/career connection is by far the most important area for study. Though they may be a great group to hang out with, to have a drink with, or to travel with, make sure that the influences do not hinder your career long-term. Do they like to drink too much? Are they involved in behavior that would not look good in the front page of the local paper? Are they content followers? On the other hand, are they focused on excellence? Are they sticklers for detail? Do they have wholesome families and off-duty activities? Do they treat you like family? The people that you surround yourself with will influence the decisions that you make and the decisions that people make about you. Know them, know what they are about and, above all, know yourself and what you are about.

For those who are interested in leadership, keep on studying. Getting that first promotion is about more than simply taking a test. You know people who took the test, but do not really know the subject that they are being tested on; the question is, do you want to become one of them? Leadership is not a science, it's an art, and you should never stop studying leadership if you want to be an effective leader. Studying leadership is not just about the classic leadership theories, it's about accumulating knowledge, skill, and traits that cause things to happen consistent with your goals and the goals of the group or organization. Study not only what works, but study just as hard about what does not work. You may know what worked for the last group in terms of leadership; the question is, will that work with the current group?

Spend as much or more time studying indicators of the future as you do studying the past. Life gives you two choices. You can spend time trying to fix the past or creating a better future. Those who might suspect that trying to create a better future is an excuse to simply disregard the past might want to reconsider. It would not be a good idea. The reality is that it takes a balance. You need to know where you have been and where you are to make an informed decision about where you need to go next. This requires studying the world around you as much or more than you study the circumstances you are in. You might have a great idea, might be doing really well, and may think you are on your way, until the world changes. When the world changes around you, it's the information you gathered about the coming change that helps you be ready for it after it arrives.

Read books, take classes, talk to people, attend meetings, go online, get you certificates, and complete your formal education. Study, learn, and gather as much information as you can. You never know it all, and what you don't know could kill your career, could kill others, or could kill you. Your studies can also be the fountain of abundance not only for your career, but for your life. It's okay to keep studying.

Peter Demontreux

We Were on Fire. Literally on Fire.

I got to know Pete by email and then by phone as I was writing my "Close Calls" column about him in *Firehouse* magazine. While all firefighters should train and prepare for the day when they must take "the risk of their career," Pete was trained and prepared to take that huge risk—and he saved a life. He carried out one of the most stunning rescues in recent memory. Pete found a man trapped in the smoke and confusion in the rear of a burning Brooklyn brownstone—and then pulled him literally through a wall of fire. "We were on fire," Demontreux remembered from the August 2010 fire. "I could feel the burns. I could feel my face burning." He received the Chief Ray Downey Courage and Valor Award from Fire Engineering, and FDNY's highest awards for bravery: the 2011 James Gordon Bennett Medal and the 2012 Dr. Harry M. Archer Medal. Pete set the example nationally by heroically saving a life, while reminding the public and elected officials to take a close look at all the good that you and we do every day.

Peter Demontreux

Things I've Learned

I stepped off the rig, grabbed my tools, and looked at the building to see what my options were to get to the rear. The chief yelled to the chauffeur of 132 that an occupant was at the front window. So now I went back to help the chauffeur in the front. The chauffeur positioned the ladder perfectly at the windowsill, and as soon as it stopped I went up. When I got to the top the first occupant was kind of in and out of the smoke. I yelled to him and then saw part of his arm in the smoke. I grabbed him and pulled him out onto the ladder. He said his friend was still inside. That's it. I pushed him down into the rungs and went over him. The chauffeur took him down and I masked up and went in. I did a right-handed search until I got to the back of the front room. It was very hot and literally zero visibility. I could start to feel the side of my face tingling through my hood. I must have run into a table or some kind of piece of furniture, and I remember thinking this is not good. It's getting hot as hell and if something comes down on me I'm going to be in trouble. I was worried maybe a china closet or something like that would topple onto me. It wasn't a good feeling, so I turned around and went back to the window I came in.

I came to the windowsill and there was another firefighter there, and I told him that it's hot as hell in there but I can hear a guy yelling. He said let's take some of these windows to let some heat out. He vented with his hook and I went back in. I went along the left side this time, and at some point I turned my head up and could see some flames above my head. I kept going along the left and somehow funneled through a tight kitchen to the back room where the guy was leaning out the window hollering. I looked out the window for a fire escape but here was none. So I took this guy's arm and held it under my left arm and led him back to the front, to the ladder. We made it to the threshold between the kitchen and the front room and at some point the whole room was bright orange. It was like someone turned the lights on. I had the man's arm and pulled and ran as fast as I could across the room to the window I came in. At this point we were both engulfed in fire. I threw him out onto the ladder to another firefighter waiting on the tip. Then I dove out onto the ladder and we were hit with a hoseline from the front stoop of the building.

Some things I've learned through all of this are:
- It is important to drop two lines in front of a fire building when you're an engine chauffeur who has to fly away from the building to

get to a hydrant. The engine that night did that, and the second line was the one that put out the flames that were on the victim and me. That heads-up move saved both of us further injuries by stopping the burning quickly.

- I had what people who write fire procedures call a "confirmed life hazard." The first occupant told me about his friend, and also I heard him screaming. These things made me more confident to keep going on with the search under these conditions.

- I wasn't searching to find the fire. I had my hood up. I wear it up most times anyway at fires. If I didn't, there would be no way I'd be able to tolerate searching in those conditions. I was burning through it with it on correctly.

- You can't beat an experienced and calm firefighter working alongside you when you-know-what hits the fan. The truck chauffeur that night was an experienced firefighter and chauffeur. Any words exchanged between the two of us were said in a calm and professional manner, which I think is something that is huge as far as setting a tempo at a job.

- The most important thing that I've learned from this is that any number of things could have gone wrong that night and led to a complete disaster. It's a humbling fact. I could have tripped up on the way out, something could have fallen on top of me while searching, someone could have moved the ladder mistakenly, anything. Something totally out of our control could screw us up at any fire.

Ty Dickerson

From the "Sackroom" to the Fireground

Chief Ty Dickerson served as the chief of the College Park Volunteer Fire Department in Prince Georges County, Maryland. The CPVFD, located on the University of Maryland campus, is unique in many ways, but primarily because they have firefighters who "live" in the firehouse. That's where Ty started and climbed the fire service ladder—as a "live in," or locally known as a "sackroom" firefighter. Located directly above the apparatus bays, the sackroom is designed like a campus dormitory. In exchange for their housing each semester, 18 full-time student volunteers work three to four evening duty shifts per week. Each student volunteer makes several hundred responses per semester, serving in all capacities from firefighter/EMT, to apparatus driver, to line and staff officers. In the fire protection field his entire career, Ty now serves as the chief of Lexington, VA, Fire-EMS. His years of unique experiences, both operationally and wading through the "politics" of the job, provide us with some valuable insight.

Ty Dickerson

The Role of the Company Officer

- The company officer is an integral member of the department: the leader, supervisor, and participant in all department functions.
- The company officer establishes a positive station environment, making it a pleasant place to live, work, and play.
- The company officer follows, communicates, and enforces the rules and regulations of the department, serving as an example of the type of individual of which the department can be proud.
- The company officer is responsible for the safety and welfare of the members at all times, especially during emergency operations.
- The company officer evaluates the effectiveness of the operational forces and maintains the standards established by the department.
- The company officer makes all decisions based on the good of the department. The success of our operations is based on how we operate as a team, as a company, and as a department.

SPECIFIC COMMENTS RELATING TO THE ROLE OF THE COMPANY OFFICER

Setting an example

In order to develop the proper attitude in members, it is essential that we as officers act the way we want members to act. This means that we must follow the rules that we set and, more importantly, that we live up to the standards that we want to maintain.

Station staffing

Because we do not rely on a fixed system of staffing the station at all times, it is necessary that all officers monitor the personnel situation. At no time should the officers allow the available personnel to reach or go below the prescribed minimums (six) without attempting to correct the problem.

Apparatus staffing

Each officer is expected to ensure that the apparatus is properly staffed based on the existing overall staffing situation at the time and the nature of the incident at hand. Minimum staffing goals remain in effect at all times (two on an ambulance and four on engine). This is considered only the minimum acceptable level of staffing. It is our goal to send maximum crews whenever possible; however, good judgment, actual staffing limitations, and the need to staff multiple apparatus at one time should prevail.

Policies vs. procedures

It is important that we all understand the difference between policies and procedures. A *policy* is statement of the department's position on an issue under normal circumstances. It is intended to be used for guidance in a particular situation. A policy also is designed to be flexible so that it can be adjusted when necessary. A *procedure* is a fixed set of steps to guide an individual to a specific end result. It is intended to provide an exact step-by-step guide to accomplish something in a specific situation.

Most of our rules and regulations, standard operating guidelines (SOGs), and so on are issued as a policy statement. This is so we as officers can use our experience and judgment to solve our problems in an efficient and reasonable manner. We continue to operate in this mode so that we are not locked into ridiculous situations and so that the department can grow and change with the times.

Treatment of personnel

Our major enemy in maintaining an adequate number of members is competition with other activities that are social in nature. It is therefore necessary for us to make the fire station the most desirable place for our members to spend their spare time. We should do what we can to make it fun to hang out at the firehouse. Be considerate of needs while remaining consistent with department goals or policies. Treat members the way you want to be treated.

Forming of attitudes in probationary members

This is basically a reiteration of the thoughts in paragraph one. The stress here, however, is to make sure that we start the process of attitude development as soon as a person joins as a new probationary member. Between the officers and the training committee we must support the mature and responsible members as early in their careers as possible and work to improve those who cannot or will not meet our standards.

Safety and welfare of members

It is our responsibility to minimize the occurrence and severity of injuries to our members and ourselves. Safety is always our overriding concern in the station and during emergency and nonemergency situations. Any time someone is performing an unsafe act, we must take immediate steps to correct the situation. This rule or guideline remains in effect at all times. We must also look out for members during extended operations or during adverse weather conditions. There is an old U.S. Cavalry maxim that goes:

> *The horses eat first.*
> *The troops eat second.*
> *The officers eat last.*

Follow this rule at all times. Being an officer is a position of sacrifice, not privilege. Think of your position as one of servant leadership.

Enforcement of rules

Unfortunately, it is necessary to have fixed rules and regulations for the self-directed members who think they can do what they want whenever they want. We have to see to it that specific and important rules are uniformly applied and enforced. The members (and officers) signed an agreement to obey our rules and regulations. This means that they volunteer to do what is required of them. The first and most disasterous breakdown in a fire department is the degradation of standards. We must not let this happen. We should do everything humanly possible to maintain a quality organization.

REQUIREMENTS FOR THE COMPANY OFFICER

Minimum qualifications
- Any requirements found in the constitution and bylaws
- Requirement of city or county code and ordinance or personnel regulations
- Two years of fire service experience
- Satisfactory completion of the Company Officer Training Program
- Satisfactory completion of the classroom work for the Pump Operator School and the Ground School
- Attendance of at least 50% of the regular department meetings and drills for the previous six months

Required knowledge, skills, and abilities
The candidate for company officer must demonstrate the following:
- A thorough knowledge of the organization, functions, policies, rules, regulations, and procedures of the fire department
- A thorough knowledge of fire suppression, rescue and EMS principles, practices, apparatus, and equipment for single company operations
- A thorough knowledge of basic personnel management principles
- A working knowledge of the functions and activities of other departments, organizations, and associations as related to fire department operations
- An understanding of the geographic, climatic, and economic characteristics of the departments response as related to fire service activities, such as planning and operations
- A thorough knowledge of the local and state laws governing functions and activities of the fire department under normal, emergency, and disaster conditions
- An ability to communicate thoughts and desires to others under your command and to your supervisors
- An ability to drive one engine in the department upon appointment and the understanding that officers should be able to drive all department apparatus

Dave Dodson

All Smoke and No Mirrors—What We Do See Can Kill Us

For years we would hear units arriving on a scene with "smoke showing" . . . and it was just smoke. Dave Dodson helped us make a lot more sense out of what was "just" smoke. The man behind The Art of Reading Smoke, Chief Dave Dodson is a 30-year veteran who helped us understand that smoke issuing from a building is the only clue to predict fire behavior. He has taught us that we must "read" smoke before choosing attack tactical priorities. While a nationally recognized speaker and author, Dave is about as down-to-earth as anyone can get, and he always takes the time to answer our questions. Smoke explosions, backdrafts, flashovers, light or dark smoke, drifting or fast moving aggressive smoke. Damn, it used to be so simple. Thanks to Dave, he has made it so that we can understand what we are getting into.

Post-fire site visits can help firefighters understand how the fire spread and how the building began to fail from fire and heat effects.

What You Don't Know Can Kill You

Looking back at 25 years of fire suppression duty—and the lessons learned—is quite humbling. When I look back, I can honestly say that I was more "lucky than good" at too many fires. In today's fire suppression world (explosive smoke and lightweight buildings), firefighters must be "good"— they can't afford to be lucky. It is with that thought that I'd like to share a personal journey that might help you discover what you don't know and give you some personalized motivation to pursue solutions.

The journey begins with my childhood desire to be a good firefighter. I basically grew up in the firehouse and was always amazed at what the firefighters knew and the incredible things they did (the widely popular NBC TV series *Emergency!* provided further fuel for my desire). I wanted to not only be like them, but be one of the best of them. Right out of high school, I joined the U.S. Air Force as a firefighter. Skip ahead to my second structure fire (I was 19 years old). My partner and I were assigned to do a quick search of a second-floor day room in a three-story dorm building while the engine crew prepared for attack. We hit the zero-visibility room and searched half of it before the heat drove us out. We barely escaped a flashover; the zero-visibility room lit up just as we exited the room. The engine crew hit it seconds later. After, we laughed at our melted helmets, discolored Nomex, and minor burn blisters. At my next "real" fire, three of us on a hose crew were dropped by a collapsing floor and stairway. No injuries. We dismissed the event as just an acceptable danger of the job and shared a collective laugh at our good luck. These events, however, shook something inside me. I felt that the training I had received was very introductory, and that I had much to learn. Specifically, I discovered that I did not understand the relationship of fire behavior and building construction—and that lack of knowledge could kill me! I needed to do something if I was going to be "good."

Back in high school, my trades experience was limited to working as a clerk in a small hardware store. Short of building a few childhood forts and tree houses, I had no building construction knowledge or experience. My firefighter academy spent all of two hours on building construction. Many fire officers of that time came from the building trades (carpenters, electricians, plumbers), and it was merely assumed that everyone knew how buildings were constructed. While I learned from these officers, I felt I needed more in-depth study.

Frank Brannigan was the go-to fire service educator for building construction. My first exposure to the ol' professor was at FDIC (Fire Department Instructors Conference) in Cincinnati—a day-long class on steel buildings. I was all ears. The class confirmed two things: First, I had much to learn about fire effects on building construction; I was *way* behind in my knowledge, and the good professor motivated me to make a never-ending knowledge quest to understand buildings. Second, I realized that my in-house fire department–provided training could never suffice. I had to invest my own time, money, and energy if I were to become a good fire officer. Frank has left us, but he did leave a legacy that is forever imprinted.

The second part of my journey occurred after my USAF stint when I got a job with the Parker Fire District outside of Denver. Our chief, Duncan Wilke, encouraged training and rewarded project efforts with outside training trips. Chief Wilke sent me to a class taught by John Mittendorf, battalion chief for Los Angeles Fire. In that class (Truck Company Ops), Chief Mittendorf mentioned how important it was to "watch the smoke—it's telling you the future." My curiosity was tweaked. I wanted to learn all I could about what the smoke was saying.

As my career evolved, I used many sources to study building construction and fire behavior. I went to college for my fire science degree, bought dozens of books, subscribed to all the trade magazines, went to more local and national classes (many on my own time—no OT), and did site visitations. Although the books and classes were awesome, I have to say that the site visits became the most valuable tool in learning about buildings and fire behavior. As a company officer, I got myself in trouble for skipping daily assignments because I spontaneously stopped by a construction site and got sucked into learning the particulars of the building. At nonfire incidents (EMS runs, alarm activations, and so on), I would take care of the customer then use the opportunity to check out a few building particulars when appropriate.

If another shift or fire department got a structure fire, I would go visit the site and look through the building and see how it reacted. Interesting to note here, I never had an issue with investigators or owner/occupants giving permission. My dad was a fire investigator (Arvada Fire District, Colorado, and later as a private) so I would go help him muck on my off-duty days just for the opportunity to see how the building reacted. At one of his jobs, I took on the task of sketching a recreation of how an undocumented building addition (that collapsed during the fire) contributed to the loss. A structural engineer reviewed the sketches and taught me about shape, strength, and heat effects. The sketches were admitted as evidence in a civil suit.

The site visits also included riding time with other fire departments. I have never been employed by a fire department that had lots of fires. If my journey to becoming a really good firefighter was to be realized, I needed to supplement my educational pursuit with more actual fires. I rode with FDNY, Chicago, LA, and several other fire departments that did some firefighting on a regular basis. The experienced fire officers I met along the way shared many of the puzzle pieces that helped me develop an ability to read smoke and read buildings.

As a training/safety officer, I continued to visit sites and borrow samples of new materials to see how they reacted under load in our burn-house. When we got an acquired structure to burn for training, I would continually watch the smoke and monitor the structure in between all the fun stuff. What fabulous learning! When assigned as the incident safety officer at working fires, I would pay particular attention to the smoke—and discovered that Chief Mittendorf was right! The smoke told a story that was so much more accurate than watching flames.

Now that I'm off the street, my journey to become a really good firefighter is over, but my learning continues as I try to fill the role as a traveling fire service educator. I'm not sure if I ever became that great firefighter, but I know that I have never stopped learning how to be better. Given that, let me share a few gems that might, in some way, help you be that great firefighter:

- The best way to learn about buildings is to get out of the firehouse and walk through your community's buildings. Know your buildings!
- Turbulent smoke that fills a compartment means flashover is imminent.
- If the smoke you are crawling through is moving faster than you can, you need to cool the box and/or escape.
- It is okay to open the nozzle and cool the overhead smoke (keep the combination on straight stream—heat seeks cold).
- Read the NIST Technical Reports on fire behavior—cover to cover.
- In response to those that claim a building collapsed without warning during a fire, Brannigan said: "The warning is the brain—in your ability to understand buildings and anticipate how they will react to a fire." Never stop learning about buildings.

- Your FD won't give you all the training you need to be good; invest in yourself (and your crew).
- Go to FDIC at least once in your fire service life. If your department can't fund it, save up, work some trades, or barter favors with the family to make it happen. You'll be amazed at what you will discover there.
- "Lightweight buildings are disposable—firefighters are not." —Brannigan
- Question what you know, and never pass up an opportunity to test that knowledge.

At your next structure fire, read the building and read the smoke *before* you commit—and remember: Don't just *be* safe; take it to the next level and make it *more* safe.

Chuck Downey and Joseph Downey

Carrying on a Sacred Family Tradition in the Name of Their Dad

As you have probably figured out by now, I have deep respect for so many people who have influenced our lives, and I have even pulled this book together to let you meet some of those good folks. You have also hopefully figured out that all royalties from this book are being donated 50/50 between the NFFF and the Chief Ray Downey Scholarship. Ray, who was nicknamed "God" because of his amazing command presence, personality, and expertise in fire rescue operations, was murdered on 9/11. Of his surviving family members, two of his sons are now FDNY chiefs: Chuck and Joe Downey, following in his footsteps. "I know it's impossible to be him," says Chuck Downey, a battalion chief, "but I want to do things that he's going to look down upon and say, 'you know, I'm proud of my son doing that.'" Including Ray Downey, Special Operations Command (SOC) lost 95 men that day—totaling 1,600 years of experience.

I expect many of you reading this book were not firefighters on 9/11 and may not understand who Ray was. Now you will. After serving with the U.S. Marine Corps in the Middle East, Chief Downey was appointed to the FDNY on April 7, 1962. In August of 2001, he was placed in charge of all SOC operations—rescue, squad, hazmat, and marine—and promoted to deputy chief. Chief Downey's phenomenal 39-year career with the FDNY was built upon success after success and rescue after rescue. One of the most, if not the most, decorated men in the department, Chief Downey received five individual medals for valor and 16 unit citations. Additionally, he was awarded the Administration Medal in 1995 for his efforts on the bunker gear program and interim quartermaster system. Chief Raymond Downey's fire department accomplishments are legendary and monumental. Below is a summary of his many achievements:

- Deputy chief of Special Operations Command, FDNY
- Panel member of the presidential committee on terrorism known as the Gilmore Commission, which has been assessing domestic response capabilities for terrorism involving weapons of mass destruction
- Task force leader for the New York City Urban Search and Rescue Team, which responds to disasters both around the country and within New York State

- Task force leader for national disasters, whose team responded to the Oklahoma City Bombing, Hurricane Marilyn, and the bombing at the Atlanta Olympics
- Author of the book *The Rescue Company* and a series of videos on collapse operations
- Recipient of the Crystal Apple Award issued by Mayor Giuliani on July 23, 2001
- Chief of rescue operations, FDNY, during the World Trade Center bombing in 1993
- Team leader in response to Hurricanes Hugo, Andres, Fran, Marilyn, and Opal
- National Wrestling Hall of Fame (Medal of Courage, awarded in June, 2002)

Chuck and Joe Downey have accomplished so much in their careers—there is no doubt that their Dad is smiling down upon them. Please excuse me if I didn't take time to tell you about Joe and Chuck, as I wanted to make sure you knew their dad.

A father and son talking after a 2nd-alarm fire on Hull St. in Brooklyn, NY. Behind (then) Captain Downey is former Chief of Dept. John J. O'Rourke.

Communications

As I sit around the firehouse kitchen table for a drill, meal, or change of tours, these are just some of the most common topics of discussion: communication, probies, safety, tactics, EEO, OT, downtown-headquarters, flow paths, VEIS, and more. Whether initiated by a firefighter or officer, the drill discussion often includes the fact that there will be changes coming down. Change can be good or bad depending on the topic, but it is always good if it is going to save the lives of civilians and firefighters. One of the most common topics that constantly needs reinforcing is communications. This topic alone is one of the most critical actions at any fire/emergency operation.

Proper, brief, and accurate communications ensures professionalism and greatly reduces the chance of accidents or injuries. Both ladder and engine operations run more efficiently with proper communications.

The following are a few of the most common communication examples from experience.

FIRE LOCATION

"L127 OV to Battalion 50. We have fire showing from the second floor rear window on the 2/3 corner." As an incident commander (IC), this transmission is complete and concise. Depending on the type of structure/occupancy, I can evaluate and if needed implement any tactical changes. If this transmission were "We have fire in the rear," it is neither proper nor specific and, as a form of communication, it is much less useful information without the details of floor/side/fire/smoke, and so on.

FIRST HANDLINE

"Engine 2 to Engine 1, are you good?" When arriving as the second engine company in the FDNY, general duties are to assist with the first handline and act as the water resource officer for the first handline. A proper transmission in this scenario would be "Engine 2 to Engine 1 ECC, do you have a good hydrant" and/or "Engine 2 to Engine 1, do you need more line?" Again, making reference to specifics helps efficiency and promotes professionalism.

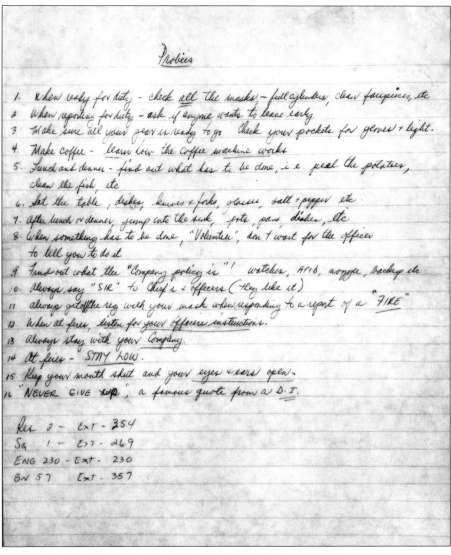

Dad's handwritten 16-point probie duties given to Chuck after graduation

APPARATUS POSITIONING

As an IC you may arrive at the scene of a fire and witness the first engine in front of the fire building with no hydrant and a water relay necessary because the second ladder came into the block from the opposite direction. "Engine 1 chaffeur to Ladder 2, I am taking that hydrant at the end of the block. Do not come into the block." In this situation the ECC knows that the second truck will block him from getting to that hydrant, if Ladder 2 drives into the block and past the hydrant.

I have one very good ECC who at times adds to the statement by saying, "Ladder 2 don't even ****ing think about pulling past that hydrant."

I have been guided by many great officers and senior members, including my brother. But no one has provided as much insight to me as my father. Having the opportunity to work with him at fires and in the firehouse and commute to work and talk fires with him has instilled the importance of proper communications throughout my career. As my dad told me when I was a probie, "On this job, you will learn something new every day, and when you think you know it all, it's time to leave."

Currently, we need to plant the seeds for those probies and newly promoted officers to lay the foundation for the next generation so that they know how critically vital proper, concise communication is to any fire and/or emergency operation.

As I was told when first handed a handie-talkie, "Take a few seconds to think about what you are going to transmit; be brief, be concise, and provide a mental picture."

Be safe.

A New York Legacy

Having grown up in a fire department family with one of the most decorated and respected fire officers of the FDNY, one would believe I knew a lot more then I did about firefighting. We were never privy to the heroic stories of our father who was known for his unparalleled leadership abilities and dedication to the job he loved. Raymond M. Downey responded to thousands of fires and emergencies during his stellar 39-year career before he was killed on September 11, 2001.

Ray Downey set the bar high for himself and those who worked for him. He expected 110% each and every time the bell went off and led by example. Ray Downey was a visionary and was involved with numerous projects that made the FDNY better during his career with the FDNY. He brought technical rescue to the FDNY rescue companies when fire duty began to slow down and emergencies were on the rise. He wrote the book *The Rescue Company*, which is still used by many who want to build an efficient technical rescue company.

When the federal government needed help in establishing an urban search and rescue system to assist local emergency responders for large-scale disasters, they reached out to Ray Downey, who was affectionately known as the Master of Disaster.

Ray Downey became the first National Task Force Leader representative for the 25 urban search and rescue teams assembled throughout the country. He led this group until he was killed at the WTC attacks and is known as the godfather of the present 28-team FEMA Urban Search and Rescue system. He responded to numerous hurricanes, earthquakes, and was the operations chief at the Oklahoma City bombing. In New York City Ray Downey started the Special Operations Command Rescue School, which now is recognized for their premier training for technical rescue incidents. Special Operations Command companies respond to hundreds of technical rescue emergencies each year. Ray Downey has touched thousands of lives with his articles and lectures and has made the FDNY and our country better for the work he has accomplished. Many have tried to emulate Ray Downey because they believed in him and his values. Because others have learned and trained with Ray Downey, many lives have been saved and his vision continues.

Joseph Downey

Downey family at Gracie Mansion for dinner with Mayor Giuliani in honor of Dad for his service to NYC and the FDNY: (left to right) Chuck Downey, Kathy Downey Ugalde, Marie Downey Tortorici, Rosalie Downey, Ray Downey, Sr., Ray Downey, and Joe Downey

There was never much discussion about the FDNY in the Downey household until my college years when my father demanded we get an education first and take the fire department test as an option if things didn't work out after receiving our college diplomas. I didn't have much time after graduating from Hofstra University to choose a career, because two months later the FDNY called. My father was thrilled for me and provided some words of advice upon entering probationary firefighting school. He was a man of a few words, but when he spoke you listened, especially about a career he was so passionate about. He said, "Make sure you always step up in the firehouse, and don't wait for the senior man or company officer to ask you to do it."

After graduating probationary firefighting school I didn't have much say on what company I would be assigned to. My father felt strongly about firehouse tradition, and I was sent to company lead by a respected and experienced captain Ed Higgins. This company, Engine 227, The King of the Hill, happened to be located down the block from Rescue Company 2, my father's firehouse, which provided me an opportunity to work at the same fires as him and also to commute together to work. Carpooling with my father and his chauffeur John Barbagallo to work was a great learning experience because it gave us time together to discuss something that was not spoken often at home. It also helped me understand how people with more than 25 years fighting fires in the busiest FDNY companies still loved going to work every tour and felt privileged to be FDNY firefighters. Having worked down the block from my father in Brownsville, Brooklyn, and eventually working side-by-side with him in the Special Operations Command gave me an opportunity to learn from the very best.

I was never given special treatment except the one tour I was detailed to Rescue Company 2 from my Company Squad 1, and he gave me the can position. This meant I would have the most sought-after position and would work right alongside him no matter what fire or emergency we would encounter that night tour. Nobody outside of Rescue Company 2 would ever get the can position, but nobody would ever question Ray Downey on why he gave it to a fireman from Squad Company 1. I saw firsthand how talented a firefighter my father was. Rescue Company 2 responded to a fire in the Sunset Park section of Brooklyn that night. As we pulled into the block a second alarm was transmitted, and fire was on multiple floors of a five-story new law tenement building.

The battalion chief ordered us to the upper floors to search for occupants due to the fact that it was 4:00 in the morning. As we tried to enter the building there were several companies on the stairs trying to advance, but it was bottle necked. My father, without saying a word, went to find an alternate way in the building. We went to the rear; the fire escapes were loaded with firefighters trying to advance in, but heavy fire was showing. We then found a ladder, and he ordered me to throw it up to access the floors above. There were reports of the stairs being torched transmitted over the radio and there would be a delay in getting to the floors above. All this time my father said nothing to me, and I just followed, making sure I didn't fall behind, because he was moving along at a good pace. We started our search on the upper floors, and as we worked our way down to the third-floor apartments my father pulled me out of the stairway landing into an apartment. We continued our search in the apartment when the interior stairs from roof to first floor collapsed.

We didn't make any amazing rescues that night, but I learned a few valuable lessons working with my father. A couple of these lessons have stuck with me, and I have tried to use them during my career. When your first choice doesn't work, you need to have a plan B and make sure you have a plan C and D in case there are other obstacles. As we all know, nothing goes as planned in firefighting. Don't get caught up in the moment; keep calm especially when things are getting crazy. Make sure you are aware of your surroundings, and you must always have situational awareness because we are dealing with life and death situations every day. Fortunately no one was hurt in the collapse, and I was able to share a special night in the firehouse that most will never experience. Being the can man in Rescue Company 2 with your father.

I have been blessed to be a member of the FDNY for 28 years and worked in the Special Operations Command for 24 of those years. I have worked with, for, and alongside some of the best in the FDNY. I have had the pleasure to work with many of the great people who shaped the Special Operations Command and who were killed on Sept. 11, 2011. Over 100 members of the Special Operations Command lost their lives trying to save civilians. Those of us who survived because we were not working or were fortunate by getting out owe it to those who sacrificed their lives to continue the tradition. It was important to get the FDNY and Special Operations Command back to the high level of excellence my father and others expected. We could not let them down. This motivated many of us to rebuild in the memory of those we lost.

Twelve years later the FDNY and Special Operations Command is as strong as ever and continues to provides the public the best service. This is displayed everyday but couldn't have been more evident than when Hurricane Sandy hit NYC October 29, 2012.

Ray Downey receiving the Crystal Apple award at Gracie Mansion from Mayor Giuliani on July 23, 2001

Hundreds of lives were saved because FDNY members used their swiftwater and water operations training to tackle something that was never experienced in NYC. We had historic flooding and moving water along hundreds of miles of shoreline. My father and the other 342 firefighters killed on 9/11 couldn't have been more proud of what they saw. We have and will continue the outstanding tradition of the FDNY for those who sacrificed their lives in the line of duty and because this is what we love to do. I was told working for the FDNY was a great job, and my father was right.

"We Shall Never Forget."

Michael M. Dugan

A Gentle Giant Who Carries a Big Stick

Mike recently retired as the captain of FDNY Ladder Company 123 in Crown Heights, Brooklyn. Not that he wanted to, but that's how it works sometimes. Fortunately, his teaching and writing continues on. While he had a very colorful and successful career while assigned as a firefighter in Ladder Company 43 in Spanish Harlem, Mike received the James Gordon Bennett Medal in 1992 and the Dr. Harry M. Archer Medal in 1993, the FDNY's highest award for bravery. The James Gordon Bennett Medal was established in 1869 and for years it was the sole decoration awarded for valor in the FDNY. As a result of its seniority among medals, it is awarded annually for the most outstanding act of heroism. The Dr. Harry M. Archer Medal is given only to a James Gordon Bennett Medal winner once every three years. A huge figure of a man, Mike stands tall, is proud of his daughters, and loves being a firefighter—one who has gone out of his way for others while usually carrying a tool in his hands.

Mike Dugan and the members of Ladder Company 123 on the roof of a fire building critiquing the fire and drilling to get better

Don't Be a Wheelbarrow

The first thing I was told by my captain, Danny Marshall, when I was transferred to Ladder 43 in Spanish Harlem, was "Don't be a wheelbarrow." I asked him what he meant by that, and he said, "A wheelbarrow will do a lot of work, but *always* has to be pushed in the right direction." A wheelbarrow is a great tool but never gets anything done on its own. As a firefighter, if you always have to be told what to do, you are a wheelbarrow. You might get the job done, but you never do more than expected or take an action on your own. A wheelbarrow rusts and rots away if someone does not always move it.

In the fire service this analogy means to new young firefighters that it is better to take an action or make a move at a fire than do nothing. The fire service needs self-starters and men and women who want to be involved and make a difference. This is done through actions; your words may speak volumes, but if your actions don't follow those words then you are just a lot of hot air. Most fire officers, chief officers, and senior firefighters would rather have someone who steps up and makes mistakes rather than not do anything.

We have made mistakes in the fire service. It's a fact of life that you make a decision based on the best information you have at the time about the current situation. Anyone who tells you that have not made a mistake at a fire is either a liar or a fool, but either way they are not someone you want to be learning from. Bosses would rather have a firefighter who tried and failed than one who never leaves the recliner unless ordered.

If you step up and mess up, at least you tried and you have learned a lesson. Lessons are very important in the fire service. If you see mistakes made and learn from them, you are getting smarter. A wise man once said, "A smart man learns from his mistakes, and a wiser man learns from the mistakes of others." After you learn a lesson, you should not make the same mistake again, and you should pass your knowledge on to your brother and sister firefighters.

The other thing you need in the fire service is to be a student of the game. You should always be learning and improving yourself professionally. By being a student you can learn lessons from other people not even in your organization or department. Finding out what is happening in the fire service can make you a person in your department that people come to for information. Someone who is informed can become the weather person for his or her company or department. Informed people tend to know the current climate in the fire service and what is new, hot, or even dangerous to your troops. They become informal leaders within their organizations. The more you know and the better informed you are, the more people look to you for advice. Read books, magazines, reports, and websites to learn what's happening. My dad use to say that "readers are leaders." Be informed and know what is happening in your profession. Remember that professionalism has nothing to do with pay or size of your department. It is based on how you and those around you conduct themselves at a scene.

Remember that your name and your brother and sister firefighters' names are on the side of your apparatus along with the name of your department or company. Always do your best to honor and uphold the traditions that made your company great, and do your best to make it better.

GLENN A. GAINES

Walk Softly and Carry a Big Pike Pole

The title I picked for Glenn is the best way I can describe him. While walking softly and carrying a big pike pole isn't exactly what he does these days, it describes him. He is a very effective, quiet, and powerful proven leader. However, when he speaks, listen carefully—his decades of experience are our benefit. We've been friends since serving together in the Washington, D.C., Metropo Fire Chiefs' Committee, as well as being neighboring chiefs.

Glenn has been working at the federal level since 2001, and since 2009 as the nation's number two "fire chief" since his retirement as the fire chief of Fairfax County, VA. During his illustrious 35-year career in Virginia, he served in numerous capacities, including fire marshal, chief training officer, and chief of operations, culminating in his appointment as fire chief, serving from August 1991 until December 1998. He was in charge of the nationally recognized Fairfax County USAR team that frequently deployed throughout the United States as well as internationally. Chief Gaines holds a degree in fire administration and has authored a fire service text and numerous articles.

In 2010, Glenn was awarded the International Association of Fire Chiefs Metropolitan Fire Chiefs President's Award of Distinction, and then again in 2011 he was awarded the Metropolitan Fire Chiefs Lifetime Achievement Award. Also in 2011, Chief Gaines was presented with the prestigious IAFC President's Award for meritorious service to the fire service. With all he has done and been through, here is just a little bit of what he has to pass on.

The Core of the Matter in Fire Service Excellence

What are the core values that provide the pillars that sustain top-performing organizations? Although we tend to look at fire departments in terms of fire stations, apparatus, and equipment, these material objects would not exist without the people of the organization. In high-performing organizations it is people who plan, organize, procure, and use these important material resources with expertise, dignity, and respect for others. It is important to recognize that each member plays a role in determining how well an organization functions and meets the demands placed upon it. Regardless of the position in the department, individuals meet or are observed by hundreds of people at work, in neighborhoods, and in social settings. Each member plays the role of ambassador for his or her department. I do not know the postmaster general of the U.S. Post Office. The Post Office I know is the young man who stops by the house to drop off the mail and gives us a big wave and smile.

If we want our organization to be held in high esteem, we must all be good citizens, perform as professionals, and treat our citizens with respect. Let there be no mistake about it, every role is critical to the success of an organization. For an example, a firefighter and an EMT respond to the home of a rather distraught father with a seven-year-old girl who has a lacerated hand, and they perform routine duties: explaining what they are doing as they treat the daughter, reassuring the father that they will care for her on the way to the hospital and that she will be fine, so drive carefully. The dad is forever grateful, yet this incident goes in the books as just another call for help. This incident represents possibly the first and last contact this family will ever experience with the department.

It is important to establish, embrace, and reinforce core values for members of the department. Some suggested core value statements follow:

- Do the routine things perfectly every time.
- Members will attain and maintain the minimum skills, abilities, and knowledge to perform the duties and responsibilities of their assigned positions.

- Each member must maintain physical and mental fitness to preform assigned duties and responsibilities of the position.
- Each member must respect other members and citizens regardless of gender, race, political belief, or religion.
- Officers must expect and demand that subordinates continually meet minimum established standards.
- Subordinates must be treated fairly and with respect by their supervisors.
- Each member must respect and treat others' property with care in the workplace and while preforming duties.
- Each member should be willing to lend a helping hand to coworkers when they are in need of assistance in completing their assigned duties.
- Fire and rescue department leadership must aggressively pursue excellence in customer service through innovative policies and programs.
- Fire and rescue department leadership must provide an atmosphere that promotes self-reliance in their officers through continuing education and training. This commitment by senior management will allow for decisions to be made at the lowest level in the organization.
- Fire and rescue department leadership must ensure that the organization is operating at a reasonable cost to the citizens, while continuing to improve services.
- Fire and rescue department leadership must ensure that personnel are provided a safe, healthy, non-hostile, and productive work environment.
- Fire and rescue department leadership must provide personnel with the facilities, apparatus, and equipment that meet contemporary service demand.
- Field officers must ensure that members do not cause unnecessary damage to a citizen's property while operating at their residence or place of business.
- All officers must set a good example for their subordinates.

The core values I have listed are what set high-performing organizations apart from other fire and rescue organizations. Each of the items listed provides for excellence in service and a department that is prepared for the next century.

Chief Glenn A. Gaines

Gordon Graham

Perhaps the Best Fire Service Instructor Who Has Never Turned Out to a Fire.

I met Gordon Graham in the early 90s when we were both "on the road" doing training. He was one of the original "members" of "The Secret List" (go to www.FirefighterCloseCalls.com and click on "The Secret List" to sign up), and one day he asked me, "Why don't you have a website?"

Well, at the time I had no money to spend on a website—and I had little time. His response was swift: you find the time and I'll fund the project. A few months later—after thinking about his offer, www.FireFighterCloseCalls.com was born. Through years of work by some great folks such as Brian Kazmierczak, Forest Reeder, Barey Furey, Rudy Horist, Ignatius Kapalczynski, Chris Shimer, and Pat Kenny, more than 4 million folks go to the worlds most visited firefighter survival website—thanks to Gordon Graham. As a part of our venture on FFCC, Gordon and I have become close friends. Who knew that a firefighter and a motorcycle cop could get along—it's a good thing that he sees things my way.

As you may or may not know, Gordon's decades as a cop (retiring as a commander with the CHP) along with his education as a risk manager and experience as a practicing attorney, coupled with his extensive background in law enforcement, have allowed him to rapidly become recognized as a leading professional speaker in both private and public sector organizations with multiple areas of expertise.

For those of you who are into motorcycles, Gordon attended the University of Southern California in their Institute of Safety and Systems Management program. He will quickly tell you that this was the best education he ever received from the best and the brightest people in the field. His professors included Harry Hurt. His relationship with Professor Hurt led to his being selected as a team member collecting data for the *Hurt Report*. Published in 1980, this report on motorcycle fatalities was and is recognized as the single greatest treatise on motorcycle safety.

A friend to the fire service, he has instructed tens of thousands of firefighters and fire officers in risk management. Recognized numerous times, Gordon was awarded the IAFC's Presidential Award for Excellence in 2005 for his lifelong work in improving firefighter safety and performance.

In 2002, Gordon became a founder of Lexipol, a company designed to standardize policy, procedure, and training in public safety operations. Today, nearly 1,000 agencies use the Lexipol Knowledge Management System, and nearly half of the states are now using this approach to law enforcement operations. Most recently, Lexipol has established Lexipol Fire—a program for fire departments to institute proven and realistic fire service policy and training systems long-term.

Gordon Graham

Predictable Is Preventable

So are you committed to risk management? "Sure we are—we are committed to risk management." RAH! RAH! With the exception of some of the nonsense coming out of D.C., these nine words are one of the biggest lies in the public safety profession. Frankly, too many public safety agencies have *no clue* on what *real* risk management is all about. And to avoid being accused of being "anti" public safety—remember that public safety is a small part of life in America—and this whole country has no clue on what *real* risk management is all about.

We don't teach it in grade school, we don't teach it in high school, and unless you have taken a specialized course at a university, you have not been formally exposed to this wonderful discipline. And hear this loud and clear. We hire young women and men and put them into the high-risk jobs of firefighting, law enforcement, corrections, telecommunications, lifeguarding, and EMS operations—and can you show me any training at the start of their career in risk management?

When I have these conversations with people one on one, they will tell me that they have a "safety" program. Good for you! Safety is part of risk management—but it is much bigger than the safety stuff. It is much bigger than the code enforcement stuff. It is bigger than the insurance stuff. Everything that gets done in every public safety agency involves a level of risk. Since we don't teach anyone about the discipline, too many people—including leaders who should know better—are getting worked up on the wrong stuff.

On the fire side of things, we spend so much time on firefighting skills—and yet the number-one cause of death is cardiovascular related. When you study the data put out by the USFA and NFPA, too many of these "heart" events involve grossly overweight personnel, heavy smokers, and people with pre-existing heart problems. Now there is a commitment to risk management.

On the cop side of things, we tend to focus a lot on cops getting murdered. Yet historically traffic collisions kill more cops than felony suspects do. On the paramedic side of things, I can show you departments around America that have EMS personnel running from call to call to call for 24 hours—and these meds are fatigued and make stupid mistakes with patients and in vehicle operations. Yep, we are committed to risk management and we are not even addressing the fatigue issue.

If anyone who reads this piece has the "juice" to do something about this—we need to have a minimum four-hour class for everyone in public safety at the start of their career to get them to understand that there is a discipline known as "risk management," and it works. There are so many things that we can do if we are "risk wise" to reduce the deaths, injuries, indictments, embarrassments, and the other nasty consequences that occur because of our lack of knowledge regarding this discipline. Risk management requires training. Training. Training. Training. *Every day* is a training day! Solid, Realistic, Ongoing, Verifiable, Training. SROVT!

Look, nearly every "serious" regional airline accident over the past 10 years involved at least one pilot who had previously failed a proficiency test. Each of these incidents was predictable and preventable. If your pilot can't pass the test, then maybe he shouldn't fly the plane. What initial and regularly scheduled proficiency tests does your department have?

We all remember US Airways Capt. Chesley "Sully" Sullenberger, who landed his airplane in New York City's Hudson River after several birds flew into the craft's engines, rendering them inoperable. Sullenberger is a shining example of one of my seven rules of risk management: training has to be constant and rigorous. Sullenberger said in an interview shortly after his heroic actions saved the lives of everyone aboard Flight 1549 that he tried, throughout his flying career, to make small deposits each day into his memory bank, knowing that one day he would "have to make a massive withdrawal." It was a sound strategy, because doing so enabled him to make instantaneous, life-and-death decisions on that fateful day. It's a lesson especially adaptable to firefighters, who make such decisions on a daily basis.

You will run into the unthinkable event someday, and you will have to make instantaneous decisions. Whether you are prepared to do so is up to you.

Please keep these truisms in mind: your fire/EMS organization must strive for continuous improvement in their personnel:

- Your department—career or volunteer—must hire quality people. If you hire stupid people, they are not going to get better over time.
- Your department's supervisors must spot problems before they become tragedies.
- Your department and its members must have a healthy respect for the dangers and risks they face.
- Your department must establish performance metrics for its personnel and hold them accountable. Rules without enforcement are just nice words.

- Your department and its personnel must be able and willing to learn from their mistakes.

Concerning my final rule, there was a woman I encountered while a member of the California Highway Patrol. The woman lived near Malibu, Calif. On three separate occasions, each roughly a decade apart, wildfire destroyed the woman's home, which she promptly rebuilt on the same spot each time.

During the most recent wildfire, I received numerous emails from people who had attended one of my lectures at some point over the years and were now concerned that I might be in danger—as I live in California. The emails, which came from all over the country, were so numerous that I eventually was forced to craft a blanket response, which I wrote firm in the knowledge that "California catches on fire every year." I wrote, "Risk management is not a class I teach; it's a way of life. Do you really think I'd build my [freaking] house in the [freaking] woods?" Predictable and preventable.

So that is my pet peeve, and my message to you: knowing things may go wrong and doing nothing about it. I am sure that there will be a lot of other pieces in this book in which various writers are venting about something—but each of their issues might be resolved if we really understood what risk management is all about.

Bill Gustin

Energy Crisis? There Is No Energy Crisis When This Man Is Around

Energy. I mean seriously, I thought I had a lot of energy—and then I started hanging around Bill Gustin. Ever been to one of his classroom or hands on training programs? Incredible. High speed. Nonstop energy and enthusiasm for the job. Don't miss the opportunity. And hold on tight. Bill's a 40+ veteran of the fire service and a captain with Miami-Dade (FL). He began his fire service career in the Chicago area and conducts firefighting training programs in the United States, Canada, and the Caribbean. He is a lead instructor in his department's officer training program, is a marine firefighting instructor, and has conducted forcible entry training for local and federal law enforcement agencies. Like many of those who have contributed to this book, he is a contributing editor and an editorial advisory board member for *Fire Engineering* and FDIC as well as writing for *Firehouse* and many other publications. Need a boost? Feeling low? Forget energy drinks—read his pieces, watch his videos, go to his website, or spend some time with the man—who has never slowed down passing good stuff on to us all.

Firefighters dig through bricks to reach Bill's dad who rotated basket out from under the parapet moments before it collapsed. Note the rope that the two firefighters slid to get down from the basket.

Wear Your Chinstrap

What is the difference between a fire hat and a fire helmet? It's the chinstrap, which is designed to keep a firefighter's helmet on his head. Unfortunately, many firefighters fasten their chinstraps across the back brim of their fire hats. There is a big city fire department that must have a written or unwritten rule that prohibits the use of chinstraps. Watch videos or look at pictures of this department operating at a fire and you'll invariably see a firefighter working on a roof without a helmet. Firefighters love their helmets to look beat up because a "salty" helmet is a badge of honor that tells the world that you've fought a lot of fire. I have to wonder, however, when I see firefighters with "seasoned" helmets with chinstraps fastened across the back brim: is the helmet beat up because it's seen a lot of fire, or does it look that way because it has fallen off their heads so many times?

When I entered the fire service in 1973, every firefighter that I knew, except my dad, wore a Cairns leather helmet. In those days, no one wore a chinstrap because old leather helmets were sized to fit your head like a glove; there was no OSHA regulation or NFPA standard requiring impact caps, adjustment ratchets, or chinstraps. Leather helmets in those days rode low on the head and were extremely comfortable. Today's fire helmets, whether leather or plastic, derive their protection from an impact cap; the outer shell is primarily a façade. Impact caps in modern helmets give them a much higher profile, a higher center of gravity, and a greater tendency to fall off the heads of firefighters who do not use their chinstraps.

I gained a deep appreciation for helmet chinstraps from my dad, a lieutenant in the Chicago Fire Department. As I mentioned, my dad did not wear a leather helmet when I became a firefighter in 1973. In 1962, my dad was a lieutenant on Snorkel 1, which was the first articulating boom apparatus in the fire service. The rig, built on a 1958 GMC chassis, was purchased from the Parks Department, where it was used to trim trees. Shortly after transferring from Engine 77 to Snorkel 1, my dad retired his old worn-out leather helmet and replaced it with a MSA "top guard" made of a thick plastic composite. At that time, the helmet was worn by firefighters in the Los Angeles area, but most Chicago firefighters preferred leather helmets. My dad's decision to purchase a plastic helmet was purely economic: the plastic helmet was less expensive than a leather one. The fact that the plastic helmet was equipped with a chinstrap was not a deciding factor. He did, however, wear the chinstrap against the advice from "experts."

Firefighters can get advice on just about anything from "experts" in the firehouse kitchen, whether they want it or not. There, over a cup of coffee, you'll get advice from fellow firefighters on a wide variety of subjects, from financial difficulties to marital problems; no subject is outside their realm of expertise. The "experts" advised my dad not to use the chinstrap on his new helmet because, if he should fall through a floor, the helmet may hang up on something and the chinstrap could strangle him. I thank God that my dad did not take the expert's advice, because that chinstrap saved his life.

In May of 1962 Snorkel 1 was operating at a "2-11" alarm fire at a furniture warehouse on Chicago's West Side. The Snorkel was positioned in a narrow alley at the rear of the building, and two firefighters were directing a heavy stream overhead into the cockloft from its basket spotted at the bottom of a second-floor window. My dad seldom operated the Snorkel from its basket, except to relieve his guys in bitterly cold weather. He always felt his job was to stand back and get a big picture of the basket, the boom's elbow, overhead wires, and the condition of the fire building—things that may not be readily observed from the basket. When a large crack developed in the parapet above where the basket was positioned, my dad and other firefighters in the alley yelled to the firefighters in the Snorkel's basket to move it, but they could not hear the warnings. Remember, this is in the day before portable radios and turntable-to-basket intercoms. My dad was a humble guy, so it wasn't until I was a young man and had a chance to talk to the two firefighters in the basket that I found out what really happened. This is what they told me: As the crack in the parapet grew and collapse was imminent, every firefighter in that narrow alley ran out of the collapse zone—except one, my dad; he ran in the opposite direction, toward the Snorkel. The firefighters in the basket told me that my dad jumped up to the turntable and operated the controls, which override the basket's controls, to swing it out from below the parapet moments before the parapet and a large section of wall collapsed. They both credit my dad for saving their lives. The apparatus and my dad were buried under large pile of bricks, a utility pole, electric transformers, and energized wires that were in the collapse zone.

My dad was critically injured and was not expected to live. He sustained multiple fractures and severe internal injuries. Then, after a few days in the hospital, doctors discovered that he had a deep electrical burn on the inside of his thigh, a result of the falling electric wires. Miraculously, my dad survived his injuries and returned to the Snorkel. He could have retired on a disability, but he loved the job and couldn't wait to return to full duty. He went on to complete his 33-year fire service career and enjoyed 20 years of his pension.

I want to pass on something that my dad told me early in my career to every firefighter who thinks it's fashionable to fasten their chinstraps to the back brim of their fire hats: "Wear your chinstrap; If I hadn't, the first brick would have knocked my helmet off and the next one would have killed me."

Turntable controls where Bill's dad was standing when the wall collapsed.

ALEXANDER HAGAN

A Truck Company Street Leader to a Fire Officer's Labor Leader

In New York City there are several unions representing the members of the FDNY, including firefighters, officers, and EMS personnel. The Uniformed Fire Officers Association (UFOA), Local 854 I.A.F.F. AFL-CIO, represents 2,500 lieutenants, captains, battalion chiefs, deputy chiefs, supervising fire marshals, and medical officers of FDNY. Captain Al Hagan is their president. I got to know Al when we ended up teaching together at FDIC a number of years ago. It was not planned, but due to one of our instructors being unable to make it, Al stepped in. A veteran fire officer (and not yet a labor leader), Al did a phenomenal job teaching the troops with a mix of humor . . . and dead seriousness. His decades of experience on the streets as well as teaching at "The Rock" (FDNY's Academy) provided a welcome gift to those in attendance. Serving as president of the UFOA, he has taken on some very tough issues: staffing, firehouse shut downs, dispatching, response times—just to name a few—all focused on what's best for his members and the citizens of NYC. A truly unique gentleman, character, leader and fire officer—Al has so much to offer—and we're thankful he eagerly participated in this book project.

Alexander Hagan

The Floor Above

I'm not sure when it started, but when I became a member of the FDNY in 1973, there was already a program called Adaptive Response, known to the troops as "the AR."

In the FDNY, generally the first-arriving units (engine and truck) go to the fire floor while the second-due units go to the floor above the fire. Back in 1973 the fire responses of the FDNY were *off the charts*. There simply were not enough units to go around . . . so the AR was born.

There were thousands of false alarms being transmitted every year. The FDNY had to adapt and provide a way to respond to every alarm with at least enough resources to begin effective operations. The Adaptive Response was a way to avoid sending the second-due ladder company until it was determined that the alarm was for a working fire.

The Adaptive Response provided two firefighters (in addition to the normal complement of five plus an officer) to trucks from 1500 hrs until midnight (the time period when the FDNY was receiving the most false alarms). Not all trucks were part of the AR program. These two firefighters were tasked with performing the duties that would have normally been performed by the second-due truck on the floor above. They operated with *no* supervision.

One afternoon I was hired (on overtime) to be part of the AR team. I had only a few months of experience as a firefighter . . . in the engine. My partner was an equally junior engineman.

Engine 36 was an active company located at 1849 Park Avenue in Spanish Harlem. My AR tour was around the corner on 125th Street in Tower Ladder 14. The chauffeur of Ladder 14 that afternoon was a guy from Ladder 30 (133rd Street and Lenox Avenue), Lou Tessio. Lou was one of the most respected firefighters of his day . . . at least in Harlem.

He asked what I was doing in L-14. "I'm working the AR," I naively responded. Lou was amazed that the job would hire such a very green guy for such a dangerous job. He went to the battalion to argue that my partner and I were not ready to do the AR. The battalion aide told Lou we were the only two guys that he could find who were willing to work.

Lou sat us down and gave us a piece of advice that served me well for the 35 years I worked in Harlem and the South Bronx: "When you pass the fire (without a charged hoseline) you become a victim," he said. Then he asked, "What does every victim want?" My partner knew the answer, "A way out!"

Tessio advised us that when we went up to the floor above that night to find the way out (fire escape) before we did anything else. After that, he told us we should look for the other victims (besides us) who might be up there and take them with us to *the way out.*

Bobby Halton

Endless Energy and Faith . . . On Behalf of Us All

Over the years I have told *Fire Engineering* editor Bobby Halton that I have suffered numerous aneurysms trying to fully understand his thoughts. Why? He is one of the smartest people I know in our business, and his research, education, intensity, and depth produce some fascinating stuff—far beyond where most of us are. Bobby pushes us into areas of discomfort because, without that, so many of us would be standing still. He makes us think outside the, er, uh . . . forget the box, he just makes us think way, *way* out there so we can be better firefighters, officers, and chiefs. Prior to his current role as editor of *Fire Engineering* and the host of FDIC in 2005, Bobby served as fire chief of Coppell, Texas, and prior to that he was deputy fire chief of operations in Albuquerque, New Mexico. Bobby started his career in Albuquerque in 1984 and rose through the ranks, holding every rank including paramedic. His primary interests are his family and *you*, the street firefighter and fire officer. Bobby is an old friend, as he is to many of you, and once you meet him, it's a lifelong relationship. There are few people with the intensity that forces us to think like Bobby does . . . and we are certainly fortunate, even if an aspirin is required every once in a while.

Have the Courage of Your Convictions and Humility of Our Mission

When first approached by my friend Billy to add to this collection of writings, I wondered what value my voice could potentially have among such influential people and incredibly gifted writers as the men and women Billy has asked to contribute to this book. I felt anything I might add would be insignificant by comparison, and then I remembered the most important advice I ever received concerning writing: write about what you know, write about what you do, write about what you're passionate about. Reflecting on that sound advice, I thought I would write about my current responsibility and duty in the fire service, and that is to opine.

It's interesting today to see how people react to one another's opinions. There is a growing adoption of a very cruel and brutal system designed to minimize and ridicule people who disagree with your position or ideology that was developed by a community organizer named Saul Alinsky and captured in his book *Rules for Radicals*. I think it's a horrible and damaging methodology. The premise is to isolate, minimalize, ridicule, and discredit the other person's opinion or position to elevate your own. I think there could be no system of debate or discourse that is more unproductive, more ungentlemanly and ungentlewomanly, or more unappealing than that which is put forth in *Rules for Radicals*. For example, rule number 12 states, "Pick the target, freeze it, personalize it, and polarize it. Cut off the support network and isolate the target from sympathy. Go after people and not institutions; people hurt faster than institutions. (This is cruel but very effective. Direct, personalized criticism and ridicule work.)"

Bobby Halton

There is a great need for us to share our opinions to have intense and passionate discussions about the issues and concerns that face us as a profession. For within the fire service as an institution, we have many issues of great importance that we need to discuss, and to have those discussions we must be willing to hear the opinions of others, try as best we can to understand those opinions, and then respectfully state our concerns with them or our opposition to them in a way that is not offensive or minimizing or abusive. The issues we need to discuss have to do with tactics, culture, politics, and a wide variety of social and professional issues that span everything "from soup (what's for dinner) to nuts (some of our favorite people)." We also need to be mindful that the fire service is not a homogeneous or monolithic profession. Among us are many diverse cultures, many religious backgrounds, and many social differences.

In my travels I have been fortunate enough to meet conservative firefighters, liberal firefighters, black firefighters, white firefighters, male firefighters, female firefighters, gay firefighters, straight firefighters, union and nonunion firefighters, volunteer and career firefighters, retired firefighters, and Republican, Libertarian, and Democrat firefighters. There is no one single model that defines the American firefighter. We are truly a reflection of that great American melting pot that began many years before our nation was even founded. Our diversity is the basis of our strength; coupled with our local foundations, we have consistently been the answers to our communities' local emergency needs—answers that, as firefighters, we have been honored to deliver with compassion, humility, and grace, answers that we found by debating respectfully and sharing our opinions courageously with one another.

What compels us is the mission, what unifies us are our cultural values, and they are and they have never changed—*loyalty, honor, duty, respect, integrity, personal courage, selfless sacrifice*. By staying true to our values and always putting the mission first, we can discuss virtually any topic without becoming rude or insensitive, without becoming abusive or cruel, and without becoming—in a word—unbecoming. Always try to see the value in someone's opinion no matter how offensive you might find it. You will never lose their respect for you, which allows them to hear the value in your position.

In his essay "On Liberty," John Stuart Mill wrote, "The peculiar evil of silencing the expression of an opinion is, that it is robbing the human race; posterity as well as the existing generation; those who dissent from the opinion, still more than those who hold it. If the opinion is right, they are deprived of the opportunity of exchanging error for truth, if wrong, they lose, what is almost as great a benefit, the clearer perception and the livelier the impression of truth, produced by its collision with error."

John Stuart Mill was a smart guy. He always held the belief that others' opinions, especially those that had been held for a long time, meant they must have some truth in them and that the folks who hold these beliefs should not be seen negatively but as people who just continue to recognize a spectrum of that truth or opinion.

He recognized that new information might render older opinions limited but nonetheless that truth, limited though it may, be must still be acknowledged and never isolated, ridiculed, or diminished. We can pick up bad habits from bad people either consciously or unconsciously. Many believe that the ends justify the means; they do not. Firefighters have always held that the means should have the same value and honor as the ends, always; anything else is just excusing evil. We should embrace those who still see different truths; as limited as we may feel they are, they are still to be acknowledged and recognized.

Firefighters must have the courage to oppose those behaviors and opinions that do not reflect our values, recklessness, bullying, and thoughtlessness. Every firefighter has the duty to express opinions, every firefighter should have the personal courage to express his or her beliefs, every firefighter should respect the opinions of others, every firefighter should have the loyalty to defend the opinions of others, every firefighter should have the integrity to express opinions honestly, and every firefighter should provide that selfless service of defending the honor of those who will not be silenced.

Joanne Hayes-White

A Big City Chief of Department . . . Who Just Happens to Be a Woman

While I had met San Francisco's fire chief, Joanne Hayes-White, at some IAFC activities, it was a specific incident that helped me get to know her better. Line-of-duty deaths have connected me to several contributors to this book whom I now consider friends. In this case, it was my writing about the horrific line-of-duty deaths by flashover of SFFD Lt. Vincent A. Perez and firefighter/paramedic Anthony M. Valerio. The goal of any fire chief is to do his or her best to ensure something like this never happens. However, there is also a group of fire chiefs out there who—in spite of having to deal with these horrible deaths—step up to lead the department forward and affect changes as required so it never happens again, and so that the members are never forgotten. Joanne and her team have done just that. I strongly encourage you to read the entire report, which is available online. Joanne was sworn in as the 25th chief of the SFFD in 2004. San Francisco is now the largest urban fire department in the world with a female chief, with approximately 1,800 members under her command. She is very proud of the fact that she, either as a firefighter or officer, has worked at every one of San Francisco's 42 fire stations. Talk about experience.

Chief Joanne Hayes-White with her proudest accomplishments, her three sons: (left to right) Logan, Sean, and Riley

A Rare Opportunity

I have had the privilege of serving in the San Francisco Fire Department since April 2, 1990. It was a great fit from day one. I applaud my parents, Thomas and Patricia Hayes, for supporting my choice of career. When I approached them in the late 1980s and told them that I was interested in a career with the San Francisco Fire Department, they were a bit skeptical, but never not supportive.

The youngest of four, I was born when my parents were 40 years old. They came from a different era. They made huge sacrifices for my siblings and me. So when I told them of my desire they had three concerns:

1. It's dangerous.
2. Not many women had chosen that career.
3. We just invested in your education; this job doesn't require a degree.

Are you sure?!

I have had the good fortune to work at all SFFD fire stations and met and worked with many of our members. I know most of our members by their first names.

I have served in the ranks of firefighter, lieutenant, captain, acting battalion chief, assistant deputy chief, and now chief of department.

In January 2004, I was appointed chief of department by then Mayor Gavin Newsom, who is now lieutenant governor for the State of California. It was a bold move. Typically, chiefs of the SFFD were chosen from the senior ranks of the department—a rich reward for a laudable career. So when I was selected, at the age of 39 with 14 years in the business, it was a bit of a surprise, even to me. I was honored to be thought of, but had my own concerns with the huge decision that was ahead. It wasn't about my ability to do the job. I was confident that I could, but I too was worried about my young family (three boys, ages 10, 7, and 4 at the time) and how I could perform both roles effectively.

Looking back over the nearly 10 years of leading the SFFD, I am so glad I stepped up and into this most important role, a huge responsibility to lead and shape the SFFD into the future.

I am proud of the team of people who make up the SFFD. Quite an array of members. We are one of the most diverse departments in the country. It is value added to have a team of people that truly reflect the community we serve. Although there is a universal language when there is a fire emergency, (such as, "Everyone out!") during a medical emergency it is *so* helpful to have a member of the team who can speak the same language as the patient, obtain critical medical history, or just be someone they can relate to.

We pride ourselves on training. It is the foundation for all that we do. Our mission is to protect lives and property. Safety is a key component every time we go out the door. There are no guarantees regarding outcome given our dangerous mission, but it is my job and everyone in the SFFD's to achieve our goals and fulfill our mission while always keeping safety in mind.

I dedicate the following remarks—works of wisdom, lessons learned—to Lieutenant Vincent Perez and Firefighter/Paramedic Anthony Valerio, two valiant members of the SFFD who made the ultimate sacrifice on June 2, 2011, at a structure fire. They will never be forgotten.

- Set high standards.
- Lead by example.
- Set the tone.
- Humility in, arrogance out.
- Have respect for self/others.
- Decisiveness; gather all facts and circumstances, then make the decision and move on to the next.
- Doing the right thing, regardless of personal risk, is not always the "nice" thing to do.
- Consistency is the key.
- Be fair, firm, and friendly.
- Embrace the history/tradition of the department, but don't let it stand in the way of progress.
- Step up or step aside.
- Engage.
- Communicate.
- Listen.
- Learn something new every day.

- Teach something to someone every day.
- Strive for work/life balance.
- Define boundaries.
- Be optimistic but realistic.
- Never expect something from someone you wouldn't be willing to do yourself.
- Surround yourself with a team of people who have different strengths and don't always think alike.
- Reward/acknowledge good behavior.
- Hold people accountable.
- Encourage teamwork.
- Encourage collaboration.
- Remain curious.
- Be inquisitive; questions are good.
- Encourage creative thinking.
- Remain accessible.
- Be safety conscious.
- Don't settle.
- Don't just keep up: Lead the way.
- Celebrate breakthroughs.
- Make lists; check off when tasks get accomplished.
- Chart the course, see it through.
- Go with gut feeling . . . most times it is absolutely correct.
- Be willing to compromise, but never compromise your ethics or principles; certain things are nonnegotiable.
- Always wear your uniform with pride; the same pride you had when your badge was pinned on and you took the oath of allegiance.

It remains an honor and privilege to serve. I consider myself blessed to be a mother of three wonderful sons and have a supportive family and also to lead a first-class fire department in a first-class city.

CATHY HEDRICK

The Loss of Her Firefighting Son Sends Mom on a Mission

If you haven't heard Cathy Hedrick speak, then you need to. Go to YouTube if that's what it takes—but listen to what she says. The petite redhead fireball has grieved each and everyday since she (and her fire chief husband Les) lost their son Kenny in the line of duty in 1992—and while we all grieve in different ways, Cathy uses her very personal loss so that we all can learn. Cathy and Les are both active members of the National Fallen Firefighters Foundation. They are both still members at Prince George's County, Maryland, Engine Company 27 (Morningside VFD), where their son was a member. We are humbled that Cathy has offered to share some words with you in this book.

Kenneth "Kenny" Hedrick

Kenneth M. Hedrick, Morningside, Maryland, Volunteer Fire Department 1992

On January 12, 1992, Kenny was killed in the line of duty. He was a volunteer firefighter with Morningside Volunteer Fire Department in Maryland and had responded with his coworkers to a call for help from a family who lived near the fire station. Their home was on fire and they worried that a seven-year-old boy was trapped in the house. Squad 27 arrived on the scene first, before any apparatus with hoselines. Kenny and his fellow firefighters arrived on the scene and began to search the house. They found a small boy and carried him out to the paramedics. The child was pronounced dead a short time later. Returning into the house to search for more victims, without any leadership or supervision on the scene, young firefighters were left to decide what their next moves were, unsure whether they were making the right choices. Several minutes into the incident, Kenny became trapped in the basement when fire conditions deteriorated. The other firefighters later found him in the basement of the house. He had died alone that night, and that changed the lives of everyone at the scene and our family forever.

Kenny was born in Maryland. His father was a volunteer fire chief in Prince George's County, and his mother a nurse whose specialty was working with surgical patients. He and his sister spent several years as children at the firehouse where they learned about volunteerism and giving back to your community. At a very early age, probably around three years old, Kenny wanted to be a fireman and drive a big fire truck. He wanted to be like his firefighter dad, who was the fire chief of Morningside Volunteer Fire Department.

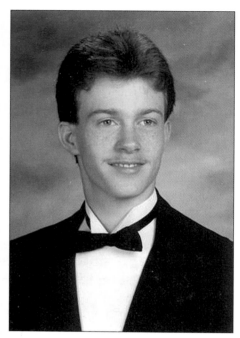

Kenneth Hedrick

When Kenny turned 16 years old he went to the fire station and completed the application to become a firefighter. It was then he began his fire service education. He earned certificates in Firefighter I & II, Hazmat Training I & II, and took many other courses and workshops. He enrolled in as many classes as he could manage while still in high school. He even managed a social life.

He was always the "social director" among his friends; they would look to him for guidance on all social matters that interested teenagers. Kenny loved family, made it a point to spend as much time with them as possible, and had a close relationship with all his cousins. He was very charismatic, but at the same time was a prankster.

Finally, Kenny graduated from high school and began submitting applications to fire departments. He was determined to become a full-time firefighter. He wanted to be the best and pursued his education. When his senior yearbook was published, there was a question asked of each senior student: "What do you fear the most?" Under his senior picture was his answer, "I fear never becoming a career firefighter."

Kenny became that career firefighter with the Prince George's County Fire Department; his badge was presented to us, his parents, posthumously.

Kenny's family has dedicated time working with the National Fallen Firefighters Foundation. We want to promote a legacy for firefighter life safety and supporting young firefighters who, like Kenny, want nothing more than to be the best fireman they can be.

The departments have made great strides in firefighter safety. But there are still many areas that need improvement. We miss our firefighter every day. The leadership let us down that night on the fireground. They had the responsibility for our son's well being. They let themselves down by not pursuing excellence in safety. Most of all they let Kenny down. He was their brother, who was very well trained but lacked experience.

My husband, Les Hedrick, was the chief of the Morningside VFD at the time of our son's death. He describes it as "a father's unimaginable worst day." I was at home with my family when we were notified of the incident when my heart was broken. Almost immediately the guilt started. What if . . . what if . . . if only . . . would he still be alive? I went through every emotion, then every possible scenario in my head after hearing the details of the event that night.

Les's advice, as a fire chief and a father, is to say to the fire service, chiefs, future officers and all firefighters *train, drill, practice*, and enforce strict accountability policies. Your young firefighters can be well trained but need your *experience* and *guidance*. Treat each firefighter as if they are your son or your daughter. You have an obligation to your families to always encourage and demand that the leaders in your department have the knowledge to ensure that *everyone goes home*. Otherwise it could be your heart that is broken, when you notify the next family.

Cathy Hedrick

Cheryl Horvath

Be What You Appear to Be

The value of "walking the walk" and not just "talking the talk" related to effective leadership is not always well known in the fire service. Cheryl Horvath, operations chief of the Northwest Fire District in Tucson, Arizona, elaborates on this truism—and lives it. This is evident in that one of her favorite quotes is from Socrates: "The shortest and surest way to live with honor in the world is to be in reality what we would appear to be." We need to understand our overall and individual "purpose" as firefighters and fire officers, then do the right thing. I got to know Cheryl years ago when she led Women in the Fire Service, now called the International Association of Women in Fire & Emergency Services (iWomen), and her solid and collaborating leadership within that organization. Cheryl is also a member of the IAFC-FRI program planning committee and collaborates with other local women firefighters to coordinate Camp Fury, a camp for girls to encourage them to seek nontraditional careers, particularly in emergency services. She serves on the board of directors of the Girl Scout Council of Southern Arizona and the Imagine Greater Tucson Leadership Council. In 2010, Chief Horvath was the recipient of the YWCA Women on the Move Award for her service to the greater Tucson community.

These women pictured with me are rock stars in the fire service. They work for Tucson Fire Department and were instrumental in helping put together the first girls' fire camp in the greater Tucson area. This picture is from the Women's Foundation annual fundraising luncheon. The Women's Foundation of Southern Arizona provided the seed money for our first girls fire camp. Pictured along with myself from right to left are Captain Diane Benson, Deputy Chief Laura Baker, and Captain Nancy Avery.

What I've Learned

Being a firefighter riding backward on an engine is a time-honored profession and the greatest job in the world. We are appreciated for everything we do, even if it is just being out in the community at the grocery store or driving our first-due areas. People wave at firefighters. They love our big red trucks and shiny equipment, even at times where we may not deserve that attention. People feel better just waving at us as we ride in our engines, and it is amazing to be a part of their appreciation.

We have long maintained, and it is a fire service tradition, that we serve all in their time of need. We take care of the homeless, the rich, the indigent, the young, and the old. Regardless of someone's economic or educational status, we will help them when they call us. We have built pride in our profession being the "first responders" to local emergencies, and we need to stay close to our core values as we serve our communities every day. It is this type of open-mindedness that helps our fire departments become closer to our communities in assessing need, risk, value, and where we can best position ourselves to be a community partner.

What I have learned is that giving your time is the most important gift you can give to your family, your service, or your community. Whether it is as union steward, an officer within the department, a volunteer on a work committee, a community volunteer, an advocate for change, or simply a busy parent, being involved is one of the most important freedoms we have. We have a great opportunity to impact our families, our departments, and our communities just by simply being more involved. People want to hear from their firefighters. They learn from what we have to say and we have much to learn from others. They automatically embrace us because they trust us; yet so few of us choose to exercise that right every day.

We cannot afford any longer to close our doors to our communities. The fiscal crisis has shown us the necessity of building cooperative relationships within emergency and local services that create ideal opportunities for efficiencies and sustainability. If we do not embrace the idea of being focused on our mission and embracing community partnerships, we will be ousted by a better business model that speaks to giving our taxpayers highly reliable and well-trained service at the lowest possible cost.

Diversifying our work force is a key to our success. We need firefighters and officers who think differently from each other; who understand the importance of asking the question "why" in the right context; who embrace and celebrate good fire service traditions while transitioning our profession to a new normal; who tolerate difference but not disrespect; who understand the importance of "team" in all work environments; and who represent our core values as we serve our communities every day. My experience has been that those "teams" whose members at first glance appear very different from each other are much better at understanding an entire problem because each person perceives a situation according to his or her own perspective and is able to add to the solution in a much more comprehensive manner. We live in an evolving society that oftentimes requires diversity of thought and complex answers.

Labor and management must come together and work in tandem in facing the fiscal, political, and philosophical threats affecting our survivability. We must challenge the status quo and learn to assess our environments together with a foundation of trust and community service. Money spent in labor-management battles is a waste of taxpayer dollars. Money spent battling harassment and discrimination lawsuits are a waste of taxpayer dollars. Money spent challenging hiring and promotional practices is a waste of taxpayer dollars. There are easier solutions to the array of issues we face in the fire service if we can all agree to stay in the discussion long enough to hear and understand each other's perspectives. It is when we walk away from problem solving that we create divisiveness and misunderstanding, and ultimately impact our future as a valued community partner. Our future depends on the actions we take today and whether we can embrace our challenges together to move forward.

This picture represents my early years in the fire service, teaching at the Illinois Fire Service Institute. As an adjunct instructor for IFSI, I had the opportunity to teach throughout the state and country, meeting and working with many great people. Although this was hard work, it was extremely rewarding to train with recruits and work shoulder-to-shoulder with some of the best instructors at the time. This experience also helped me keep my skills current and capable, a key component of being a fire officer. My ability to lead my company into emergency situations was greatly enhanced by this experience.

Otto Huber

Moving Forward—No Matter What It Takes

Chief Otto Huber is my chief of department and has mastered being a chief better than most. His focus is always on what's best for the citizens/customers, as well as what's best for his firefighters. His longtime service has allowed him to clearly understand the politics both locally among the varied elected and appointed officials and also in the fire service—which can often be splintered with bureaucratic egos that are not always focused on what's best for these two groups. While some give up or back down, that's not the case with Ott. Formally working with many area chiefs, they have formed a collaborative group that allows the important stuff to get done while saving dollars and greatly increasing services. I first met Ott in 1995, and more than once we caught up to vent over the frustration of not being able to move forward regionally as quickly as we thought best. What struck me, though, was that no matter what barriers were tossed in his way, Ott kept our department moving forward, no matter what it took. And today, he continues that mission both working within the Northeast Fire Collaborative (of six area communities with progressive and aggressive chiefs with like values) as well as the Loveland-Symmes FD.

A perfect FD? Not at all. But a very good department with excellent folks and some of the proven best firefighters and medics who ever hit the street. Spend any time at LSFD and you will see that it is a department that is always changing, evolving, training, and improving to do what's best for those dialing 9-1-1—no matter what it realistically takes. It is a pleasure to serve with Ott and our crew as a deputy chief at LSFD.

Otto Huber

Leave It Better than You Found It

My thoughts for the future of this crazy business we are in are that we need to do a better job of not being our own worst enemy. We need to learn—I mean really learn—from our mistakes and the mistakes of others in our service. We need to listen to each other with open minds, open ears, and open hearts and leave our egos at the door. This is truly the only way our service is going to propel toward the future. We have seen incredible sacrifice in our ranks over the past 25 years, some inflicted on us by war and some inflicted on us by us. By *us* I mean because we fail to learn from the past; we fail to train; and we fail to utilize best practices and allow our egos to get in the way. The fire service today has opportunities to embrace new technology new science and new practices like never before—ideas, thoughts, and results that have never been available to us in the past. Let's embrace that change, hone our craft so that we can continue to be the best profession in the world, and do it in a safer, calculated manner.

My second rant is to remember why we are here and for whom we took that pledge to protect. We get so self-absorbed at times about what heroes we are and how the taxpayers or customers need us. We need to step back from time to time to remember that we only exist for them. We are an essential service for our communities, but without them we don't exist. So let's focus our mission on them to serve them and to dedicate our professionalism to serving their needs. Finally, be a student of our profession. I have been in this fire service for more than 35 years, and I continue to be amazed every day about how much more there is to learn. Get engaged in local, state, and national committees and organizations; be a catalyst for change and improvement. We talk about being our brothers' keepers. To do that it takes putting yourself out front, getting involved, and participating in this great profession. Take the chance, get engaged, and make a difference. Leave our service in a better position than it was in when you got here.

Ron Kanterman

Love and Passion Is Icing on the Cake

Spend any time with Ron and you will be exhausted. I mean that in a good way... a real good way. Passion and love for the fire service is a common thread with each of our contributors in this book. And then there are some who are Tasmanian devilish. Ron would be among those—with that *"whoa!"* icing on the cake. Super high energy and in love with the fire service, you can't miss it after your first five seconds spent with him. I have made a habit of hanging around those kinds of firefighters. He started in 1975 with the FDNY. He left in 1989 to serve as a fire chief of one of the nation's largest industrial fire services. After retiring from that service, Ron then served as career fire chief in New London County, Connecticut. He has a bachelor's degree in fire administration and master's degrees in fire protection management and environmental science. He is also an advocate for the National Fallen Firefighters Foundation and is Chief of Operations each year at the National Fallen Firefighters Memorial Weekend ceremonies. Here is a quick glimpse into the very diverse fire service mind of Ron—who loves to pass it on.

Chief Ron Kanterman

Leadership

We've all had bosses who appear to be good leaders but are terrible managers, and vice versa. Both disciplines take hard work. Management entails lots of planning, organizing, staffing, delegating, budgeting, and other responsibilities. Can you be a good leader *and* a good manager at the same time? Or good at one and not the other? Or lousy at both? Yes to all three! Remember that leadership isn't necessarily what's on your collar. Respect for rank comes with that rank, but respect for you as a person comes with having the right stuff. Think about the best leaders, officers, and firefighters you ever worked with. What made them what they are or were? I'll guess they were trustworthy, dedicated, and well-read people with great integrity who had respect for others at the highest levels. Right? Now think about the worst leaders you've come across. You can learn from the bad ones too because you also know what *not* to do!

Consider some of the great leaders of all time. They were able to lead the masses and bring them to the place they wanted their people to be. For example, Dwight Eisenhower, Abraham Lincoln, H. Norman Schwarzkopf, and Fiorello LaGuardia. All of these leaders had one thing in common—*vision*. If you are going to be a leader *in* your organization or the leader *of* your organization, you must have a vision. Don't confuse your vision statement with the mission. Most emergency services organizations have a mission statement that include words like service, dedication, best, customer, quick, efficient, effective, ability, and so forth. But a *vision statement* is much different. It's your opportunity to shape your vision into what you believe the organization should and could look like. Put aside the budget and all the other current obstacles, and develop your vision for your organization. The leaders above were effective because they were also all *great communicators.* If you want to be an effective leader within your organization or beyond, you must have a vision, the passion to make it work, and the ability to communicate it at all times and at any and all costs. Most importantly, you must first believe in it yourself.

You must create the environment and lead by example. No apologies here—just do it. Chief Peter Lamb from North Attleboro, Massachusetts, says, "What you allow to happen without your intervention becomes *your* standard." If you continually let the tail wag the dog and the day comes when the dog must wag the tail, you will have to go over Mt. Everest to get there. You must set the stage, create the environment, set the tone, and do whatever you have to do, but you must lead at all times, not just when it's convenient. You are charged with setting the tone for ethical behavior, even if you were the biggest prankster in the firehouse or told the best dirty jokes. Once you get elevated to the next level, "You can't play cards with the guys anymore," as a former boss said when I moved up a notch a long time ago.

Remember to develop yourself functionally and technically so you can speak, operate, and lead at the proper levels across the board. You don't necessarily need to know how every new tool operates or have it in your hands when you're at the higher levels of the organization, but you need to understand the concepts so you can support the need. I can't make a 4:1 Z-rig mechanical advantage system anymore, but I know what it's for and why the rescue company needs this system to operate.

Our customers dial 9-1-1 and ask us to come and make their problem go away. The average American doesn't know or care whether we are paid or not—"I dial 9-1-1 and somebody shows up and helps me." That's the bottom line. But it goes deeper than that. As a leader, you must keep up with your town's demographics. Few communities' makeups in the country are stable; people are always moving in and out, and the ethnicities, religions, and genders change rapidly. New cultures bring new challenges for the emergency services, and as the leader, it's your job to keep up and ensure that your new customers are getting what they need. You may have to meet with community or religious leaders to better understand who they are and what they need. Approaching your constituents with respect for their traditions, culture, or religion will speak volumes and probably get the code compliance you're seeking, let alone access to facilities for familiarization tours.

Don't forget your internal customers as well—everyone in your department under your command. You need to fulfill their requests in the station as you would out on the fireground. Your people are your greatest asset—take care of them. Other customers include the other municipal agencies (e.g., the police, the department of public works, parks and recreation, etc.) Take care of them the way you would want them to take care of you when you call for assistance. Always personally support your department. If the department leadership talks negatively about it, especially in public, then what could you expect from your people? Most of us support our departments by simply wearing a marked shirt or jacket or by displaying a window sticker on our cars. Remember, however, that you are now a "marked person," and what you do affects not only you but the whole department.

Learn to think and act strategically. The first thing is that you need to know who *you* are. You can't do anything until you are comfortable with yourself and confident in your position. Once you've conquered you, then you can lead others and make the necessary changes to move your department forward. You must have your act together and believe in yourself before you can present anything to others. You must also know your department—every function, position, policy, procedure, SOP/SOG, rule, regulation, what to do, and more importantly, what *not* to do. You have to know your people. The success of every good leader I have known came from their ability to lead and have good people around them to carry out the mission. As a 20-year chief officer, I realize that most of my successes came from my deputy and battalion chiefs, line officers, and firefighters. I used to love talking to chiefs who thought they were bigger than their department members. I always had to break the bad news. "They're bigger than you, and by the way, probably much better." They never liked that. Too bad on them. Get that valuable input from your staff, look at best practices, and benchmark with your peers and professional associations. Today's fire service leadership has no excuse not be on top of current information and technology. It's right at your fingertips.

Remember that your "integrity" is everything. If you give up your integrity, you lose everything. If you lie to your people and they find you out, they will never trust you again. Some things you just can't get back. Maintain your integrity at all times. Your leadership legacy depends on it. Also part of thinking and acting strategically is *consistency* in how you handle your people when things go right or things go wrong. It's most important when things go wrong. Inconsistency can ruin a department, whether it's allowing four different shifts to operate four different ways, or preferring charges against one volunteer when two of them committed the bad act. Consistency is critical to keeping the ship not only afloat but upright, on course, and moving forward at all times. Leadership makes the world move in a positive direction, so contribute.

Good communications is the cornerstone of good leadership. You must be clear and concise to be effective. It's almost like giving fireground commands over the radio. *Almost.* Part of this is dignity and respect; yes, treat people like you would like to be treated. Talk to your people not at them. Take the high road. Even when the team manager is kicking dirt on his shoes and screaming profanity, the umpire quietly takes his hand and points to the top of the stadium indicating, "You're out of here." Not that you should throw them out, but remain calm, evaluate the problem, and quietly and effectively deal with it. Screaming matches don't work; you'll bring yourself down to lower levels where you needn't be. Show patience and courtesy even when the other person does not. Here's where your leadership skills really kick in again.

What will they say at your retirement party or your funeral? Maybe the standard answers are: he was firm but fair . . . a good husband and father . . . a good boss . . . he cared . . . we learned a lot from him . . . dedicated . . . could be trusted . . . never lied to us . . . and so on. If you think they may not say the things you want them to say, then you may have some work to do! Get going!

> "You do not lead by hitting people over the head. That's assault, not leadership."
>
> —General Dwight D. Eisenhower, President of the United States, 1953–1961.

Brian Kazmierzak

Juggling It All to Make a Difference

You may not personally know Brian, a battalion chief in Penn Township, Indiana, but if you have ever been to www.FireFighterCloseCalls.com, then Brian has definitely made an impact on you. Along with numerous other duties and roles, Brian runs the day-to-day operations of our website and has become a trusted friend. A great dad and husband, he is a highly enthusiastic doer with limited tolerance for those who just take up space. Brian is also known to tell it like it is. We sometimes need more of that. Much more. About a decade or more ago, I was brought to his area to do some firefighter survival training. I arrived with my three large bins of VHS tapes, which absolutely made sense to me back then . . . but not to Brian. After watching me insert, rewind, and fast forward tapes, he approached me after class and stated, "There is a better and easier way to do that." I doubted him for a while, but eventually trusted him to take all of my tapes and put them electronically on my laptop. Starting out as a reader and subscriber, he reached out and offered to help many years ago . . . and the rest is history. Brian works 24/7/365 with us to ensure that literally everything "behind the scenes" in delivering what I write for The Secret List, as well as FFCC, runs smoothly. Brian is a recipient of the coveted Fire Engineering/ISFSI George D. Post Instructor of the Year award, the Dana Hannon F.O.O.L.S. Instructor of the Year award, and the 2008 Indiana Fire Chiefs Training Officer of the Year award. In addition, Brian was also in the original Blue Card certified fireground command instructor course. Brian's high energy and enthusiasm help to "pass it on" to each and every firefighter he meets, works with, or instructs. Be it in good times or not the best, Brian has persevered in making this job—and his personnel—better at what we do.

Just When You Think You Have Your Plan Figured Out . . .

Life is full of ups, downs, twists, and turns. Just when you think you have all of these figured out, then you get sucker punched by life and knocked to the ground. In September 2011, this is what happened to me and my career. While in the end, it ended up being the *absolute best thing* that could have happened to me and probably saved my career, I was in no frame of mind to realize that during the weeks that followed. I was angry, upset, and almost lost my job and marriage in the end.

For the past 10 years I had served as my department's training officer and accreditation manager. I had the world by the tail, or so I thought. I worked Monday through Friday, 0800 to 1600, and had a take-home car. I was home every night and weekend, or at least that is what my job description said. I was teaching all the time, though in the end, rarely for my own department it seemed. I was there, I thought I was a part of the department, but in reality I was so far away from being a part of the department it was scary. Sure I made fire runs, I went to the big calls, but I was there as a chief officer; I wasn't a part of a company, and I got treated as such. In reality, I worked about 16 hours a day, every day. Even when I wasn't at work, I was dialed into work and blocked the rest out. I let my job control everything. My family came second; everything else came after that as long as work was happy. Due to this I had blocked my wife out of our marriage.

I turned to Facebook for stress relief, which ended up being very unhealthy in the end, and my "chats" on there is what led to my demise. Facebook is not always healthy. Look at the number of lost jobs, marriages, and so on. There are just way too many temptations, and I succumbed to them.

After a month off work and a demotion to captain, I returned to shift work. My kids had never known me on shift work, and I hadn't been on shift in about 10 years. I was not happy with this change, and my wife was especially *not happy*, and in reality, I was scared to death. However, I was welcomed back into the firehouse with open arms by a shift I didn't think I wanted to be part of . . . but I couldn't have been more wrong. These guys were fantastic; they taught me what firehouse living was all about. When I left shift, there were three of us in the firehouse, when I returned there were nine or ten guys a day in my firehouse. These guys became part of my family, like it is supposed to be. My driver, Dino, who is a gentle giant, a Samoan guy, ended up making my life great. He watched out for me, took care of me, and became a great friend. After just a couple days of being back on shift, I felt different. I felt like a firefighter again. People who I thought were my friends actually weren't; they had treated me as damaged goods, but the guys in the firehouse took care of me, respected me, and we had a wonderful time.

After six or seven months of being back on shift, I was given the opportunity to go to a neighboring department that was transitioning from part-time/volunteer to career as a shift battalion chief and training chief. Because the one thing I really missed being back on shift was not teaching and instructing each day, I jumped on it. This department had been through some rough times (four chiefs in five years) and didn't have a great reputation locally, but I looked at it as a challenge. Boy, am I glad I took that challenge. I have never been happier; I have never worked for a better chief. I have a boss who asks my opinion daily, and I get to share my knowledge that I learned at my old job (which I do have to say, my old department truly gave me a ton of skills, and I am very thankful for that). In addition, I get to help build a "new" fire department each and every day I go to work.

So almost two years later, I have never been happier. My family now comes first, I am home now more than ever before, and I truly feel like a firefighter again! So just because your life doesn't follow *your* road map, don't get down. Also, the day job and take-home car are not all they're cracked up to be. And really watch who you think your *true* friends are, because I have found mine, and they are the people that stood by me when things weren't so good, not the so-called friends who ran and hid when I was in trouble. Also, nothing good happens on Facebook. Stay off it or be careful!

Mason, Mitchell, and Brian Kazmierzak

Tony Kelleher

7,000 Runs Annually . . . 100% Volunteer, 100% of the Time

Tony has been a firefighter since 1995. He is currently the chief of the Kentland VFD, Station 33 in Prince George's County, Maryland, and a technician (driver) with the DCFD assigned to Engine Co. 4. Like many of us, from early youth his ambition was to become a firefighter. In his desire to learn, he has spent most of his life gathering knowledge and experience from many great firefighters. Like the title of this book, he learned from those who understand the importance of passing it on, and he joins us in carrying on that important tradition. I asked Tony to contribute because of the unique nature of his position, specifically being the chief of the busiest volunteer fire company in the busiest combination department in North America. A unique individual, he has proven to be an extremely effective fireground commander. His calm demeanor under the worst of times, including Maydays and other tough jobs, is well recognized. His focus on the importance of training sets the tone for so many others—and is well understood as the priority by their firefighters. He and his officers keep their members focused on what's important, and that has resulted in a well respected company—among volunteer and career personnel alike. The volunteers of Company 33 are made up of more than 50 individuals, with 15 to 20 actually living at the firehouse—and all are justifiably proud of what they do. As with a good football team, it takes a good coach and staff. Here are some of Tony's thoughts as he "passes it on" to you.

Mentorship and Motivation in the Fire Service

A LIFE OF LEARNING

As I author my contribution to this publication, I have 1,000 different things going through my mind. My imagination is running wild about the several facets of the fire service and what each of them means to me. Honestly, I'm even struggling with which topic would be best to come straight from my heart. After a long while, it has become very clear to me exactly what I want to contribute to the fire service through this book. I have decided that I want to write about being a mentor, motivation, and where that responsibility lies in every position of our great fire service.

You see, the only thing I've ever wanted to be in life was a firefighter. From the youngest age I can remember my father taking me to several firehouses and/or fire department functions (musters, flea markets, and so on). He was truly my first mentor and was the reason I developed such a passion to achieve my (fire service) dreams. Throughout my early adult years and as I progressed through the ranks, I had the opportunity to have several additional mentors, ranging from probationary firefighters to seasoned chief officers. No matter how much time they had on the job, all of them had one thing in common. Each of them cared enough to take the time to teach me, mostly about the fire service, but often about life in general. Of course this was no easy task for any of them. I'm sure it was even very frustrating, at times, for them to "get through to me," but they never stopped and certainly never gave up. As I've matured, I reflect back upon all of this and am humbled to have been given the opportunities that I had. Today I feel that it is now my responsibility to honor these men by doing the same for others.

Tony Kelleher

BEING "THAT GUY"

As odd as it may seem, a number of individuals have trouble being a mentor. Most also see motivation as an unattainable goal within their respective department(s). To be completely honest, both can be accomplished rather easily. The key is to stay motivated yourself and have the self-confidence it takes to be that mentor to someone else. No matter how much time you have in the fire service or what rank you hold, the opportunity is there for you to teach someone something about our craft or to get them excited about what future holds in store for them.

THE RESPONSIBILITY IS YOURS

The probationary firefighter

Motivation: As the new guy or girl you are being watched from the minute you walk in the firehouse door. You will be making your first impression, time and time again. Oddly enough, you already have a huge amount of responsibly on your shoulders. It is up to you to learn everything you can and build the initial foundation that will develop into your reputation. If all of this isn't enough, you also have the responsibility to learn every historical aspect of your department. Knowing where your department came from should be a huge motivation to carry on the good traditions. Last, but definitely not least, your motivation should be to accomplish everything you can throughout your fire service tenure. Never stop learning, never give up and never let someone tell you it can't be done.

Mentor: Although you have a minimal amount of time in the fire service, there will surely be others who have the same amount of experience as you do. Most likely they will look to their equal (you) for help in their time of need. Be willing to pass on the things that you learn and accept the things that they pass on to you.

The firefighter

Motivation: So, you've got some time on. The biggest thing to remember is that you must not get complacent. Complacency can injure or kill firefighters. We must remember why we do what we do. In essence, no matter how bad things can get, remember why you became a firefighter. Whatever reason you came into the fire service, keep your passion and desire burning strong. It's up to you to keep your department up to par and continue to drive forward from within.

Mentor: Routinely you will hear, "I'm just a firefighter, what positive impact can I possibly have on my department and its members?" You may not realize it, but the firefighters set the pace for everything that goes on in the firehouse. At the Kentland Volunteer Fire Department in Prince George's County, Maryland, we have a saying: "The firefighters run the floor." This means that they are the group of people that set the standard from day to day. As a firefighter, you must pass on each and everything you learn to the next person. It is always easier for a newer firefighter to approach you instead of a line/chief officer. You have to remain approachable and realize that your impact is substantial, as you are an unofficial leader.

The line (company) officer

Motivation: You are now the person who has been given the responsibility of leading a group of firefighters. One of the first things to remember is that you are *still* a firefighter. Don't let the power of this title go to your head. Your motivation should be that of the firefighters in which you supervise. Now that you are an officer, it is up to you to set the expectations. Staying motivated should certainly be one of them.

Mentor: You obviously (hopefully) were placed into your position for great reason. As an officer, everyone you supervise will be watching your every move. As was mentioned previously, don't become complacent. If you become complacent, so will your men and women. You also will soon become the friend, guidance counselor, relationship specialist, and so on. One thing that is often overlooked is that you will need to teach the firefighters you supervise how to do your job. It is up to you to make sure that everyone is ready for the next step in their volunteer or paid career.

The chief officer

Motivation: You've made it to the top, but just like the line officer you have to remember that you are still a firefighter. As with any rank, don't let the power go to your head. In fact, use it to make positive change. With that being said, being humble, picking up a broom to assist with cleaning-up the firehouse or giving a helping hand to re-rack hose certainly goes a long way. You would not believe how much this motivates the troops.

The department is your ship, and you need to set the standard on how it operates. You also have to set the mood. Remaining positive as the leader of your organization, no matter how stressful times can be, is a must. You are the person that keeps the individual and department morale as high as possible. You have the power to take your department to the next level, just as long as you value the members that will help you take it there.

Mentor: Lead by example, stay positive, stay in touch, and provide your members with nothing but the best. It is your responsibility to teach your line officers how to do your job, just as you did with the firefighters when you were a line officer. Some of the best chief officers in the history of the fire service were the most down-to-earth individuals to walk the planet. You have to remain approachable.

Firefighters from the Kentland VFD after fighting a working fire in a vacant school building. Chief Kelleher is pictured in the center with the white helmet shield.

Pat Kenny

The "Not-So-Often-Spoken-About" Losses within the Fire Service

Pat Kenny is a fire chief in Illinois. Several years ago I read about his worst day—the loss of his son to suicide. Sometimes when "bad" things are brought up, we very humanly find it easy to look the other way. That's how we as firefighters sometimes look at many issues—including the issues of health and wellness—be it toward ourselves or those we work with or love. Pat approached me several years ago with the thought that unfortunately in the fire service, discussion about close calls pertaining to mental health are not discussed. Many times the stigma of a suicide by a member of that department or a member of their family is kept hidden. The result can lead to harmful situations for the surviving family members both inside and outside the department. We don't always share the warning signs, and in some cases, even after the situation occurred. This denial leads to less likelihood that a firefighter who has experienced a close call in his or her own life, and lived to tell about it, would not feel comfortable sharing that experience. Pat is on a mission to help us understand—and change. Among many projects, Pat runs the "Personal Survival" page at www.FireFighterCloseCalls.com. He also supports and is involved with the national efforts by the IAFC and the NFFF to raise awareness and provide training to chiefs, officers, and firefighters. Perhaps most importantly, Pat has stepped up to share the loss of his son Sean—so that we can all gain a better understanding of just how real the issue of mental health can be—and how we can learn from it.

Learning a Lesson from Sean about Superman

Did you join the fire service believing you could make a difference? Did you believe you were capable of becoming a modern day superhero—cape and all? That you could do what others only dream of? Did you think that someday there might be a life saved because of an action you contributed to? You bet you did!

Do we transfer that same sense of purpose from the public we serve to members of our department—or to our family? What if you personally, as the chief officer, are struggling to protect a member of your immediate family from harm? Or you learn a department member is having family problems? We all have experienced these situations or know of those in our department who are struggling with terminal illnesses or deaths of children, spouses, or other loved ones. How do we reconcile our superhero image when we cannot protect the very ones we love?

Brotherhood is a term that lately has come to make me cringe. It looks more like an excuse to hide behind the cape than to meet the obvious head on. We tend to care for each other when it is convenient and comfortable. This includes taking care of yourself!

I challenge you to keep an open mind about where I am going.

A firefighter at the kitchen table complains of heartburn, shooting pain in his left arm, and tingling into his fingers, and there's no doubt the next move is to activate EMS. Change the situation to a firefighter who normally is engaging and the life of the party, but is now quiet, sullen, and withdrawn. Take any immediate action here? I hope so, but I bet not, in most cases. I believe the lack of action is not due to indifference, but is based on being uncomfortable with recognizing and acting upon those symptoms that could indicate a mental health challenge.

I learned firsthand the feeling of ineffectiveness after coming face to face with my inability to protect someone I loved. I lost my 20-year-old son, Sean, to suicide after almost 15 years of battling mental illness. Throughout his illness, I believed department members would have trouble understanding and accepting Sean's mental illness. How could they possibly understand the concepts and ramifications when my wife, our other two sons, and I were living in that environment every day, and we felt lost? How could they trust me to save the day when I couldn't even save my own son? So I confided in few and never offered the full truth. The irony is that if Sean had brain cancer, I would have told everyone who would listen so that both Sean and my family had the support we needed.

That decision, in hindsight, was a mistake. Wearing that cape took a toll on my family by trying to keep up this "super" facade. As fire chief, I also missed an opportunity to create an organizational culture more accepting of mental health, a huge error in judgment.

I still struggle with the feeling that I was a failure at my most important responsibility here on earth—being a father. I would trade any of my personal and professional blessings in exchange for my son's well being and one more hug. Frankly, I would trade whatever time I have left of my life for that deal.

Back to our withdrawn firefighter, "What am I supposed to do?" Or, "It's none of my business." Or, "I don't want to intrude." It becomes very apparent that we are not trained or equipped to fix these challenges, but we sure can lend some comfort and support.

During Sean's whole painful journey, I never personally sought counseling. *No, not Superman! I don't need it because there's nothing affecting me.* After all, the problem was with my son and, by God, I will make things better.

Well, I didn't make things better—not for Sean or for anyone in my department, as it applied to mental health understanding. I missed out on that golden opportunity. As I stood next to his open casket, I knew I was viewing my son for the last time on this side, and I was in the depths of a depression I never knew existed. For the very first time, I felt what he had dealt with on a daily basis for at least the last five years of his life.

I am honored Chief Goldfeder asked me to be a part of this book. It is important that you realize the point of this section is not to have you pity me or my family. You would miss the point. My passion instead is to raise awareness that just because we are firefighters, our capes do not protect us from the pain of mental illness in our firehouses or in our families. It may even be affecting you.

In this superhero analogy, I look at mental health challenges as Superman's kryptonite. It can penetrate our protective shell and weaken us. So how do we deal with the kryptonite? Better yet, how do we overcome the kryptonite to provide understanding and assistance to our fellow firefighters, their families, and ourselves?

Many states have answered the challenge of responding to firefighters in need immediately after a critical incident through a network of response teams consisting of highly dedicated, educated personnel. You need to know how to bring in those teams locally.

However, it's just the beginning. My concern speaks more to what is in place for those who suffer either from long-term effects from an incident, or when the root of the problem has nothing to do with an emergency response, as was my situation. So let's look at some lessons I learned that I hope can assist you.

One of the greatest lessons I learned from Sean's life was the power of therapy. You cannot tough it out. I turned to a therapist who knew Sean, and even more critically was familiar with the fire service. She made me face my demons of guilt and self-doubt, and that my profession and my heroic persona were going to make it more difficult to begin the healing process. She made me put my cape down and realize that I was neither Superman nor God, and both those realities were okay!

The second lesson learned was that you have an impact on the culture in your department when dealing with mental health issues, whether it's a fellow firefighter or family member. By not sharing more of my son's struggles with my department, I was sending the message that I was ashamed of this dark secret.

I am a big role-model guy, and here was an opportunity to model this progressive notion that mental health is just as important and real as physical health. Instead of seizing that opportunity, I rationalized that I was saving everyone from that pain. The reaction since my son's death has reinforced the notion that many in our profession suffer similar mental health challenges, either in their own family or in their fire service family.

If I had been more public about Sean's illness, I could have nurtured a culture that is open to mental health issues. Now I am not proposing that everyone open their private issues at roll call, but that the concern for and maintenance of the mental health side of our profession is lacking. We should understand that it is a worthy challenge to our firefighters' well being, as well as our own. Just the increase in documented cases of post-traumatic stress syndrome (PTSD) should have our antenna up.

If you do have programs, such as an employee assistance program (EAP), don't be satisfied just by their existence. As a chief officer, make sure they know about the fire service culture. They cannot help if they don't get it! In fact, they can make things worse if acronyms like SIDS and LODD are terms they first learn about in a counseling session.

Next, look into what insurance will cover. Many EAP plans limit the number of visits that are covered. Picture your firefighter, who is just making a connection only to be told, "Sorry last visit; go back to the start." That will likely be the last visit!

If you're in a volunteer organization with no insurance, what options are out there in your community? Some counselors will donate services to firefighters or reduce rates. Many hospitals have psychiatric departments, and clergy can set up a chaplain program. Remember in all cases, that no matter who offers to help, you still have the responsibility to educate them about your unique culture.

Also, be aware that the National Fallen Firefighters Foundation is leading the way with their work on Life Safety Initiative #13. Look to them for guidance.

I have prayed to take something positive away from losing my son. Sean taught me the final lesson learned and it is the most important one. Those who suffer from mental health challenges, no matter what degree, are courageous people. They are not weak. Mental illness is not a tumor you can see on an x-ray. As chief officers and firefighters alike, we must set up a system to provide help to those in need of treatment.

Now you may be thinking, "Thanks Chief for the advice, but most of what you talked about is out of my control. That's the chief's job, and I am *just* a firefighter."

Well, my firefighter friend, we are all *just* firefighters. We all can make a difference through our own actions. Be open, positive, and supportive of those who face mental health challenges.

If you think anything I've said has merit, create a culture where mental health challenges are as accepted as physical diseases. The stigma of mental illness as a weakness has to be eradicated.

I think Sean's life's mission was to teach not only our family about mental illness, but the brotherhood of the fire service, too. I believe he lost his life so others might live. So do me a favor as you live out your exciting career: Become a member of Sean's team. Make a difference as to how mental health is accepted, and embrace learning as much as you do the new suppression or rescue techniques. For that, Sean and my whole family thank you.

This picture shows my family (left to right): sons Brendan and Patrick; my wife, Eileen; myself; and Sean at our Turkey Bowl in 2000. It was the last family picture I have that includes all of us, and it captures the fact that they were, and will always be, the most important team in my life!

Stephen Kerber

Scientific Firefighting—Is It as Simple as Putting the Wet Stuff on the Red Stuff?

We as firefighters often do a wonderful job of perpetuating stories "we've heard" and over time allow them to become fact. Urban myth type stuff. Sometimes what we have heard or "know" are factual—but sometimes they are not. The funny thing is that when we actually stop to pause and find out if this or that is true—we are often surprised. Real surprised—to the point that we *still* refuse to believe it! Steve Kerber has become a well-known "prover" across the country—and he is being listened to. As the director of the UL Firefighter Safety Research Institute, Steve has led fire service research and education in the areas of ventilation, structural collapse, and fire dynamics. Prior to UL, Steve worked in the firefighting technology group at National Institute of Standards and Technology (NIST) conducting research to improve the safety and effectiveness of the fire service. He has led studies working with the fire departments of Chicago, New York, and Toledo, to name a few. While at NIST, Steve was awarded two Department of Commerce bronze medals and a gold medal for his research. Steve was also deputy chief and training officer of the College Park VFD in Prince George's County, MD. I have been lucky enough to listen to Steve speak and read almost all his work. While the science of firefighting is a "constant learn," Steve has made it a little easier for so many of us by helping us understand why—and why not. He has left me thinking about that phrase, *"If we see fire, put it out" . . . and that I haven't seen a fire get worse by putting water on it.* Things that make you go hmmm—welcome to 2014.

Steve Kerber

Understanding and Partnerships Save Lives

Never stop asking, "Why." If you are told, "Because that's the way we have always done it," do some digging. That never has been and never will be an acceptable answer. Safe and effective firefighters are those who are students of the job and realize that there is information outside of their departments. They also know that all information is not created equal. With a good foundation of fire dynamics knowledge, you can better understand what information is questionable and what experience is worth incorporating into your toolbox. Just because the fire went out does not mean that the tactics deployed during that incident were the most efficient or even correct. You have to remember that you only had one vantage point for that fire and that your actions may have combined with others actions to create the end result.

The fireground is complex, and the fire environment is ever changing. Study your fire environment and what it means to your tactics, your safety, and the community you are protecting. Changes in the fire environment may mean changes or modifications to your tactics. This does not mean that they were wrong, but things simply change and the fire service has to adapt and overcome. Simple phrases like "venting equals cooling" and "vent early vent often" have serious limitations. It is important that we understand these limitations. Limiting the air limits the growth of the fire, but applying water on the fire as quickly as possible continues to be the tactic that does the most good for life safety, incident stabilization, and property conservation.

It is important that the fire service and fire research communities continue to work together to improve firefighter safety. These partnerships have resulted in numerous tactical considerations, including the following:

- Structural collapse should always be considered in your size-up. All residential floors can collapse in your operational time frame, especially an unprotected engineered floor system.

- The stages of fire development change when a fire becomes ventilation-limited. It is common with today's fire environment to have a decay period prior to flashover, which emphasizes the importance of ventilation and its timing.

- Forcing the front door is ventilation and must be thought of as ventilation. Although forcing entry is necessary to fight the fire, it must also trigger the thought that air is being fed to the fire and the clock is ticking before either the fire gets extinguished or it grows until an untenable condition exists, jeopardizing the safety of everyone in the structure.
- No smoke showing means nothing. A common event during the experiments was that when the fire became ventilation-limited the smoke being forced out of the gaps of the houses greatly diminished or stopped all together. No smoke showing during size-up should increase awareness of the potential conditions inside.
- If you add air to the fire and don't apply water in the appropriate time frame, the fire gets larger and safety decreases. Coordination of fire attack crew is essential for a positive outcome in today's fire environment.
- After the front door is opened, attention should be given to the flow through the front door. A rapid in rush of air or a tunneling effect could indicate a ventilation-limited fire.
- During a vent, enter, search (VES) operation, primary importance should be given to closing the door to the room. This eliminates the impact of the open vent and increases tenability for potential occupants and firefighters while the smoke ventilates from the now-isolated room.
- Every new ventilation opening provides a new flow path to the fire, and vice versa. This could create very dangerous conditions when there is a ventilation-limited fire. You never want to be between where the fire is and where it wants to go without water or a door to close.
- The greatest chance of finding a victim is behind a closed door. If you get into trouble and need to escape, closing a door between you and the fire buys you valuable time.
- Fire showing does not mean the fire is vented; it means it is venting, and additional ventilation points will grow the fire if water is not applied.
- You can't push fire with water. Air entrained by a stream will cause heat to flow through a flow path. Just like any other tactic, there is a correct way to flow water into an opening such as a window or door if you do not want heat to follow the flow path.
- Sounding the floor for stability is not reliable and, therefore, should be combined with other tactics to increase safety.

- Thermal imagers may help indicate there is a basement fire but can't be used to assess structural integrity from above.
- Attacking a basement fire from a stairway places firefighters in a high-risk location due to being in the flow path of hot gases flowing up the stairs and working over the fire on a flooring system, which has the potential to collapse due to fire exposure.
- Coordinating ventilation is extremely important. Ventilating the basement creates a flow path up the stairs and out through the front door of the structure, almost doubling the speed of the hot gases and increasing temperatures of the gases to levels that could cause injury or death to a fully protected firefighter.
- Floor sag is a poor indicator of floor collapse, because it may be very difficult to determine the amount of deflection while moving through a structure.
- "Taking the lid off" does not guarantee positive results. Vertical ventilation is the most efficient type of natural ventilation. It allows the hottest gases to exit the structure; however, it also allows the most air to be entrained into the structure. Coordination of vertical ventilation must occur with fire attack, just like with horizontal ventilation.
- It is not possible to make statements about the effectiveness of ventilation unless one includes timing. Venting does not always lead to cooling; well-timed and coordinated ventilation leads to improved conditions.

Steve Kerber directs an experiment as UL, NIST, and FDNY examine fire dynamics at Governors Island, NY.

Rhoda Mae Kerr

Picking Her Battles

Perhaps you followed that issue down in Austin, Texas, a few years ago in which the fire chief wanted her members to briefly stop and make sure intersections were actually clear before proceeding through red lights, stop signs, things like that. Some firefighters were concerned about that policy ordered by Fire Chief Rhoda Mae Kerr. Under the policy, firefighters must come to a complete stop at intersections before proceeding during an emergency. They don't have to wait for the light to turn green—just make sure it is clear. Not everyone agreed. State law in Texas (like just about everywhere else) allows firefighters responding to an emergency to drive over the speed limit and to go through red lights and stop signs. *You just can't hit anyone.* And the only way to make sure that doesn't happen is by briefly stopping—and then going through when it is clear. She worked it out with her troops and they—and the public—are better protected. An example of a chief saving us from ourselves. That's why they are called chief. That's why she did and does what she does. I've known Rhoda Mae for a few decades, and she has never been shy about pushing what she believes to be right. Correct. Appropriate. She *never* fails to speak up. Maybe that's because she is a fourth-generation firefighter. She started in 1983 and is the fire chief of the AFD, served in that same position with the city of Little Rock, Arkansas, and she was also deputy fire chief at Fort Lauderdale, Florida. It's not easy being the chief . . . just ask'em. Some decide the battles aren't worth it and ignore what shouldn't be ignored. Then there are those who don't do that. Here is more from one of those, Rhoda Mae Kerr.

What I've Learned

Growing up in New Jersey, I never dreamed of becoming a firefighter; I actually dreamed of becoming a teacher. Some might think it's funny that it never crossed my mind as a kid, since I come from four generations of firefighters! A career in the fire service just wasn't an option for me in that time period. Remember, in the 1960s, women were expected to become nurses, teachers, or homemakers . . . not firefighters.

I went to school to study physical education and health. I followed this career path for 12 years, becoming a coach and physical education teacher. I have always been driven by change, so it's not surprising that one day I decided it was time for a major change. And right about that time, the possibility to become a firefighter presented itself. After some thought, I decided to pursue the opportunity because I knew I could do it! Firefighters had always been in my family, so I thought, why not? Becoming a firefighter was not my first choice in life, but I came to realize that sometimes, the second or third choice ends up being the best one. When I made my decision, I ran with it and never looked back.

My career in the fire service has challenged me, tested me, and continues to surprise me every day. It is never boring, but rather, so very fulfilling. As a woman in a predominately male field, I have faced many challenges and struggles. These struggles also apply to my position. As a fire chief, I am under the microscope even more. My vision is that we reach a point in the fire industry where gender does not matter, that we are judged only on our abilities to get the job done.

I would advise women to get as much and as broad an experience as possible within their department. Furthermore, be willing to accept new positions and move around within the organization. Also, it is essential to build relationships both internally and externally. I believe the key to success is to be engaged and visible within your community and your organization.

Education is also a vital avenue to pursue; at a minimum, a bachelor's degree should be completed. I would advise women within the organization who want to stand out among their peers to obtain their master's degree. My last note of advice would be to seek out mentors within the industry as well as in various other organizations.

Chief Kerr represents the Austin Fire Department at hundreds of events each year, including serving as a judge for the "Best Lemonade" Contest as part of Lemonade Day Austin.

I have learned to always stay true to myself. I have embraced challenges because challenges are always opportunities. No matter what obstacles I have faced in my life I have always held my head up high and maintained an upbeat outlook. Humor is important; it helps me handle all kinds of adversity. I promise it will make you feel better in any situation. I know my role as a fire chief is important to young girls striving to accomplish their dreams in life. I am an example of how you can be anything you want; never give up on your dreams. I have been dedicated, driven, and loyal throughout my career.

I am honored that I have been the first female fire chief in two large departments, and I know that my impact in that role will help open the door to that possibility for generations of women to come.

At the Austin Fire Department, we like to start our future firefighters early! Chief Kerr takes the time to show one young man what might be in store for him at our training tower.

Scott D. Kerwood

Letters after the Name: A Solid Foundation for Firefighters and Fire Officers

Years ago it was unheard of in the U.S. to see a fire officer's name signed with education-related initials following. MS, BA, OFE, EFO, PhD . . . initials like that were reserved for "professions" such as doctors, lawyers, and others who sometimes seemed to be "bragging" about their credentials. In the late 50s, a very progressive and outspoken New Hampshire fire chief named Donald Holbrook proposed to the IAFC that they establish a national program and award for those who have achieved measurable success in their careers. While it took a few decades, the IAFC formed the original Commission of Fire Accreditation and the National Fire Academy established the Executive Fire Officer program. At least two states, Ohio and Texas, have multi-year classroom and course study fire officer executive programs (patterned after the national EFO). The program in Ohio is coordinated by the Ohio Fire Chiefs with graduates identified as Ohio Fire Executives (OFE), while in Texas the Texas Fire Chiefs Association coordinates the program with graduates identified as Certified Fire Executives (CFE). One of the most respected gentlemen in today's fire service is Hutto Texas Fire Chief Scott Kerwood. Now hold on, you ready? That's Dr. Scott D. Kerwood, PhD, CFO, EFO, CFPS, CEMSO, MIFireE. Whoa! Scotty holds a bachelor's degree in engineering technology, fire protection and safety, a master's degree in public administration, and a PhD in public policy administration with public safety management and homeland security specializations. Does it matter? Well, it didn't take away his reputation as a veteran fire command officer. He still goes to fires in his area, statewide and nationally as a part of IMAT. So then, why does it matter? As a firefighter or fire officer, your bridge to the community, elected officials, and city hall is the chief. How educated do you want your chief to be? What credentials would you like your representative to have? Think about it. Scotty has some thoughts to share.

This Isn't Your Fire Chief's Fire Service Anymore! Stop Sniveling about Education

We say it all the time, but do we really believe or care what we say? I mean, we talk about education in the fire service, but then we still let individuals without the proper education (not training) elevate to the position of fire chief. We are not a *profession*, no matter how much we want to think or try to convince ourselves that we are. We are an occupation/vocation/association/ organization that plays in the arena of professionals. We are not the same as the city/county/district manager, the budget director, or board members. If we were truly a profession, we would not allow uneducated individuals to be elevated to the highest level in a fire department, no matter the size of the organization. We would require—no, demand—that these individuals making the most important decisions for our nation's fire departments and communities be on par with the bureaucrats that determine the fate of the organization. And it does not matter if you are a paid or volunteer fire chief. The issues are still the same. Your education as a fire chief must be at least equal to if not greater than the people you are dealing with. This is no longer your fire chief's fire department where you can be the most qualified firefighter based on the amount of your training or your fireground experience and get put into the lead position. Nor can you be the one who becomes the fire chief because you politically backed the right candidate, even if that is how your fire department has always done things. You are just practicing *organizational inbreeding* if that is what you are doing. Who are you kidding? Are you really making a better fire service, or are you just doing this for yourself so you can say, "Look at me, I am the man (or woman) . . ." If that is what you are doing or you did, then get out of the fire service. You have no business representing the members of your fire department in the bureaucratic or political arena as a professional or calling yourself a fire chief.

Do you ever wonder why, when you read various trade publications, it is the police chief who gets put in as the interim city manager or permanent city manager, and not the fire chief? It is their education. The police service far outruns the fire service when it comes to the education of its top executives.

Maybe that is the same reason we always see more police officers and larger police budgets while we try to scrape by on what we can get. Maybe that is the reason we see more federal dollars going to law enforcement than the fire service. I know what you are thinking: "I am here and doing fine." Well, you are lucky. More than likely, if you were to go outside your local area looking for a job as a fire chief, regardless of the size of the organization, you probably will not be given a second look without your university education. You may know how to fight fire or command an incident, but is that what you are really doing as a fire chief? Probably not. If you do not have your education, then get off of your butt and get it now! I do not care if you are in the twilight of your career. Your troops are probably laughing at you. You tell them that it is important, but you will not even walk the same walk you ask of them. Maybe to you it seems fine, but it really isn't. How much respect does that garner you? If you are not educated exactly like your budget director, your city/county/district manager, or your city/county/district attorney, then you need to leave. You are, in their eyes, just a firefighter who knows how to put water on a fire and talk about saving the blue-eyed, blond-haired little girl. And then when it comes budget time, you are not really taken seriously, even though you think you are. So although you have their respect as a firefighter, you are not respected as an equal.

So get rid of your uniform and dress like them. Move around the halls of the city/county/district administration buildings like them. I am not talking about forgetting fire service traditions. Those things are important, but that is not what this article is about. What I am talking about is you looking the part of an executive who runs an organization. If you are not educated, then you better become educated. And I do not mean an associate's or a bachelor's degree. You better have an MBA or a MPA if you want to be at their levels. There are so many ways today that you can get your education. Whether you are in the classroom or online, it does not really matter. Look around you at the various colleges and universities. Ask your city/county/district manager or board member where they went to school and then go there. Whatever you do or however you do it, just do it. Stop making excuses and whining that you do not have time, or the money, or the energy, or the desire, or that you are at the end of your career to get a degree. You just need to do it. As I said, this is not your fire chiefs' fire service anymore. So make it better than what your chief left you. Make your community, your fire department, and your family proud of you as you walk across that stage. Let your children and grandchildren know that education truly is important to you. Remember, the tassel is worth the hassle!

If you have your education, then this article is not meant for you. Please forgive my ranting about the uneducated members of our five-bugle fraternity that will not follow the same educational path that we did in order to make this truly a profession. Thank you though for leading the way in and for the fire service.

Scott D. Kerwood, Fire Chief, Hutto, Texas, Fire Department

Bob Khan

Keeping the Phoenix Rising

In 2006 Bob Khan became the fire chief of Phoenix, Arizona—a department of more than 2,000 members. Bob spent 14 years on the apparatus before beginning his climb through the ranks, and was promoted to assistant chief in 2001. He was well recognized for his work in the community, educating the public on life safety issues. One example: through an extensive public awareness campaign and the formation and implementation of the PFD Adopt-a-Fence program, they were able to reduce drowning fatalities to a two-decade low. That's a huge deal, and it takes a leader who understands the tragic outcomes to be motivated to make the change like he has. Chief Khan has been married to Peggy since 1989, and is the father of two daughters. He and Peggy adopted the girls from China and consider the experience a privilege and a blessing. So what's my connection? Knowing him from back in the day when he was PFD's PIO is one. My interest in watching any new chief take command from a highly successful legacy chief is another. Lastly? I was interested in sharing with you the words of a man who had adopted kids. I have always held people who adopt kids in high esteem. That includes Sam and Joyce Goldfeder, the two very cool people who adopted me. Talk about "pass it on" . . . wow. Here are some "pass it on" thoughts from recently retired Chief Bob Khan.

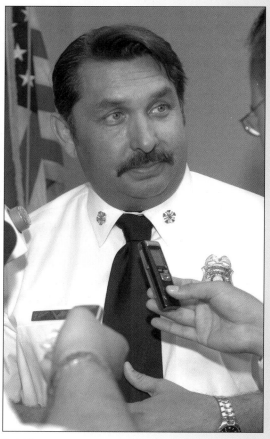

Bob Khan

The Best Job in the World

I have been fortunate enough to be a member of the Phoenix Fire Department since 1982, and I became fire chief in 2006. It's been said many times, and it is true—being a firefighter is the best job in the world. Just by doing your job, you can have a positive impact on so many people.

As you go through your career, from beginning to end, you will come in contact with people who will shape you, your approach to your work, and to life in general. A number of people have shaped my career.

Here is a summary of what I have learned from them:

Safety. The job of a firefighter can be really dangerous. Firefighters take reasonable risks to serve our customers. If you take an unreasonable risk, you put yourself, your fellow firefighters, and your customers in a bad position. Always keep the risks and the benefits of taking a risk in your mind as you work.

Professionalism. As a professional firefighter, you take care for your equipment, drive within your fire department's standards, treat each customer with respect, and wear the appropriate uniform for the work. If you want to be thought of as professional by the community, you need to act it—on and off-duty.

Teamwork. Firefighting is teamwork. Sometimes your team is small, and sometimes your team is large. Do things that build your team's ability to work together—eat as a crew, work out together, train together, and know your position. Follow the playbook when you work with other teams.

Respect. As firefighters, you must respect each other, your job, and have a great deal of respect for your customers. If you want the respect of your peers, it needs to be earned every day. The fastest way to destroy the fire service that you love is from within.

Personal development. Never stop learning. Become a better firefighter by reading fire service articles in magazines and on the web, attend classes, pay attention, get your degree, take promotional exams, teach, take care of your body and mind, and look outside of your world to understand the world seen by others.

Overconfidence. Be confident in your abilities as a firefighter but don't be arrogant. Always be humble—especially early in your career. Share what you know in a way that will be accepted by others.

Feedback. Positive feedback is the easiest to give and to receive. Pay attention to the lessons that come in negative feedback—they might make you a better person and a better firefighter.

Past practices. Remember to be innovative and accept change. Think outside the box.

Ego. As a human being, you have an ego. Fire departments have organizational egos. As a firefighter, you need to manage your individual and organizational egos. In a world that sometimes idolizes your career, you need to remember that all glory is fleeting.

Always remember those who look forward to *you* coming home.

Early in my career before becoming a chief officer, I spent 15 years in the field and worked in some of the busiest areas of Phoenix. As a firefighter and a company officer, I responded to thousands of fires and emergency medical incidents. The incident that stands out most in my mind took place when I was the department's public information officer.

The Phoenix Fire Department was in the process of fighting a very large fire in a very large pool supply warehouse that was spreading—and I'm not joking—into a chemotherapy supply warehouse. Command decided that we needed to evacuate a large area of South Phoenix, close Interstate 10 in both directions, and stop flight operations at Phoenix Sky Harbor Airport—quite a tall order.

While I prepared to conduct a press conference and deliver the news about the evacuation to the community, I made a trip into our command van to consult with the incident commander. The first person I saw in the command van was Chief Brunacini. When he saw me enter the van, he reached under his seat to retrieve something. I assumed that he was going to give me a field operations guide or some written plan for the evacuation. Instead, he handed me a Snickers bar and told me to eat the candy before doing my job—his confidence and humor at that moment were just what the doctor ordered.

The most important things that I learned from Chief Brunacini over the 25-plus years that I worked for him were to always do the right thing, be good at what you do, be nice, and be safe.

Edward Kilduff

Command This: 15,000 FFs and EMTs Making 1.3 Million Runs a Year

Imagine being in command of this: 15,000 uniformed fire officers, firefighters, EMTs and paramedics. They respond to everything from wood-frame single-family homes to high-rise structures, bridges and tunnels, large parks, and wooded areas. Add two major airports, shipping ports, and the largest subway system in the world. Piece of cake? If you spoke with FDNY's 34th chief of department, Ed Kilduff, B.A., you might think so. Cool, calm leadership. He would also make it very clear that he has the best people in the business, making his job a bit more manageable. As borough commander for Brooklyn prior to his promotion to COD, Chief Kilduff, who became a firefighter in 1977, led more than 75 units and 2,000 firefighters in the city's most populous borough. He has been cited five times for bravery, including in October 1988 when he rescued a 51-year-old woman from a burning apartment in Brooklyn, and in June of 1998 when he directed the rescue of five firefighters trapped in a building collapse on Atlantic Avenue, also in Brooklyn. Taking care of his people and being no-nonsense with respect to firefighter safety has been and remains his top priority. Chief Kilduff had firefighters and officers at all levels "pass it on" to him. He wanted to be just like them and he took that very seriously. They set an example for him to follow. Now it's our turn to learn from him.

Edward Kilduff

Get Squared Away

Every firefighter in probationary/rookie school is taught to be "squared away" and ready for the worst-case scenario on their first run. Often, our most senior firefighters and officers teach rookie school, relying on their personal experiences to validate the importance of the lesson. Safety is drilled into every component of class, and situational awareness is now the most important watchword of a rookie's life.

What we learn:

- Always be alert and ready; this is life and death.
- Full PPE at all times; no halfway or you will pay the price.
- Every run means something, if only a training event.
- Every environment we enter is potentially dangerous.
- Always evaluate risk/reward; rely on training and experience.
- Ask questions; ask questions; ask questions!
- Follow the script; no shortcuts means fewer trips to the hospital.
- Live, breathe, and eat professionalism, dedication, and commitment.
- Depend on each other, and look out for each other.
- Respect the brotherhood/sisterhood.

Then we arrive at the firehouse and begin the next part of our education. We are sometimes advised by a few "veterans":

- "We don't do it that way here; that's how they do it at the Academy."
- "This run is BS. We go there every day and find nothing."
- "I don't know why we're doing all this EMS work; I'm here to fight fires!"
- "Inspection duty is a waste of time."
- "The politicians and bean-counters are all against us."
- "Partial PPE is acceptable, because this is a nothing run."
- "Seatbelts are optional, based on convenience."

- "'What's for lunch' is more important than 'Where are we going'"?

But yet, who do you want next to you, leading you through the front door? The officer and/or firefighter who:

- Knows tactics in and out and always has a Plan B (and probably C)
- Is the first one off the rig and treats every contact with respect and compassion
- Trains every tour and not-so-subtly demands your attention and participation
- Has all PPE donned and is never surprised by a curve ball
- Communicates well and exudes pride in his/her unit and profession
- Is the role model you were told to seek out and follow

Your choice. Which member do you want to be?

- The firefighter who keeps asking questions and is always learning
- The firefighter or informal leader who steers the discussion into positive territory
- The firefighter who is proud to represent the department at a civic function?
- The firefighter/EMT who "gets it" and never loses sight that we are "here to help"
- The firefighter who greets the new members and shows them around
- The officer who is crisp in their duties and treats everyone fairly
- The officer who trains daily to make their unit better than the day before
- The firefighter who decides to take a detail to the training academy for the next class of recruits

You?

The fire service is very straightforward about its mission of saving lives and property. We have effectively enhanced our viability in the communities we serve by adapting the broader concepts of public safety agency, all-hazards capability, homeland security, and prehospital medical care into our first-responder label.

For years, I have explained to people that they are not bothering firefighters by calling them to a minor incident. I say, "Don't worry—you call, we come." It's what happens after the call is made that can make for the most thrilling, challenging, rewarding, or sorrowful day of our lives.

- Be diligent about your job.
- Understand your role on the team.
- Have integrity and be faithful to yourself and your profession.
- Make training and safety the two most critical components of being a firefighter.

And

Be the person who, when you have 20 years in the job, the new firefighter or the old battalion chiefs says of you, "That guy's squared away, I'd follow that firefighter anywhere."

WILLIAM D. KILLEN

Protecting His People, the Public, Our History . . . and Our Astronauts

One of the most interesting and humorous people who ever crawled a hallway, Bill Killen, a past president of the IAFC, is the former director of U.S. Navy Fire and Emergency Services, where he commanded and coordinated Navy shore installation fire and emergency services programs, provided technical direction and oversight of all Navy fire departments world wide for over 19 years. Prior to that, Bill commanded fire, rescue, and emergency medical operations for the Metropolitan Washington airports (Reagan National and Dulles International). Starting as a Maryland volunteer firefighter in 1956, he went on to become a career civilian firefighter with the Navy in 1960, and later joined the Kennedy Space Center fire department. He was a member of the original Apollo Astronaut Rescue team, where he continued to serve through the Skylab program. Perhaps the most personal moment that Bill and I shared together, after being friends for many years, was riding tailboard in 2002—when beloved fire service instructor Anthony "Don" Manno died. Donnie was a dear friend of mine, Bill's, and so many others—many of them contributors to this book. Following the memorial service, Donnie was loaded onto an older model pumper for his "final ride" to the Dulaney Memorial Gardens Cemetery in Timonium, Maryland. We agreed that Donnie would not make that final ride, in the hose bed, alone . . . so Bill and I rode the tailboard together with our Brother. RIP.

Bill has written many fire service technical as well as historic books, absolutely loves antique fire apparatus, and is now vice president of the National Fire Heritage Center in Emmitsburg, MD. Here he is "passing it on" once again.

Overcoming Peer Pressure in the Fire Service

When Chief Goldfeder invited me to participate in this project, I thought it would be an opportunity to share not only some of my experiences of a career that spans 57 years in volunteer, industrial, municipal, and military fire departments, but a couple of lessons learned.

What advice should I offer? If I could start over, what would I do differently?

I started chasing fire engines at the age of 10 and have been involved in the fire service since I turned 16. I can honestly say I loved every minute of it and would do it again if I could.

I don't want to clutter my comments with adjectives describing the fire service other than to say not everyone who joins the fire service is totally committed. What I mean by that is to many it is just a job, and at the end of the shift the job is not taken home. Others get so involved that their dens are shrines or museums filled with fire memorabilia, and they live, eat, and sleep fire service every waking minute. Many families proudly acknowledge fire service lineage for generations.

God blessed me with a wonderful and diverse fire service career as a volunteer while in high school, 5 years with a Navy fire department in my hometown, nearly 10 years with the Kennedy Space Center Fire Department, fire chief of a combination department in Central Florida, faculty at the Maryland Fire and Rescue Institute, fire chief for the Metropolitan Washington Airports and Director, U.S. Navy Fire and Emergency services for nearly 20 years.

There were numerous mentors and friends throughout my career, and I am proud to say I have tried my best to pass on to others the good points they taught me. This is an opportunity to share what I learned.

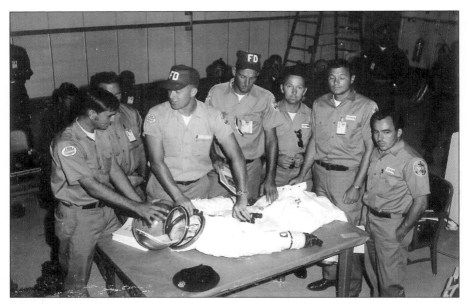

Astronaut Rescueman Bill Killen explaining the procedures for removal of helmet and oxygen lines from the Apollo astronaut space suit. Left to right are Ed Bidault, Charles Mullis, Killen, Rex Yates, Rod Hobbs, Randall Marlow, and Charles Short. (Photo credit: Bill Killen collection.)

Rookie school teaches the fundamentals needed to perform firefighter duties, and on-the-job experience hones the skills to get the job done safely. Firefighting is constantly evolving with new technology and equipment. Having observed so many changes starting on the back step of a '35 Ford American La France to buying highly technical computer-controlled fire apparatus for the US Navy 40 years later, I have seen changes first hand. I can only imagine what the fire service of the future will be.

The key to being a part of the changes to come and the implementation of new technology is going to require education and training beyond rookie school and on-the-job training (OJT). Education in the fire service is a never-ending process, and as you advance through the ranks, education is a major influence of your success. This applies to everyone from the rookie all the way up to the fire chief or fire commissioner. I regret that I didn't enroll in a fire science program sooner; it would have made a difference in promotional opportunities. Professional associations offer so much, and young firefighters need to be members and contributors locally, as well as at the state and national levels.

Astronaut Rescuman Bill Killen demonstrating the emergency release mechanism on the Apollo command module hatch. Left to right are Killen, Rex Yates, Rod Hobbs, Randall Marlow, Ed Bidault, and Charles Mullis. (Photo credit: Bill Killen collection.)

Living and working in the fire service is similar to living at home with your family. One of the biggest influences is peer pressure. Peer pressure can be positive or negative and may determine the level or degree of success.

During my tenure with the Kennedy Space Center Fire Department, I saw changes, both positive and negative. During the first three months on the job I learned enough to score high on the crew chief's and driver operator's exams. The examinations were comprehensive and challenging. I had not been on the job long enough to be eligible for a crew chief promotion.

After six months on the job, personnel voted to be represented by the Transport Workers Union who represented firefighters on the Cape side. Union negotiations provided pay raises and some benefits and seniority for promotions. The fire chief was so irritated he changed the promotional exams to a 20-question quiz, and seniority became the determining factor for promotions. Promotions outside of the union were undesirable, and very few from the bargaining unit sought them.

If opportunities arose that the union opposed or did not support, it was not a smart thing to volunteer or participate in because of peer pressure. More than a year after the Apollo One fire killed three astronauts in January 1967, NASA directed the establishment of the Astronaut Rescue Team. The company had not discussed the program with the union, and there would be no pay increase for rescue team personnel. The union, for the most part, did not support the concept, and those who volunteered for the team felt some peer pressure.

Nine of us resisted the peer pressure, and on October 6, 1968, we began training in preparation for the launch of Apollo 8. Our work schedule changed from 24 hours on 48 hours off to 8 hours per day Monday through Friday on a 40-hour workweek resulting in a pay decrease. The union reluctantly represented the Astronaut Rescue Team, and the pay issue was resolved. In 1972 as the Apollo program was winding down, a permanent astronaut rescueman position was established. I served with this team through the Apollo and Skylab programs and resigned in 1974 to accept a fire chief position with Orange County, Florida.

You might wonder why I used this experience as a lesson learned. I resisted peer pressure to be a part of something special, something more important than I as an individual or the union. I learned so much about the Saturn V launch vehicle, the Apollo Command Module, and the Lunar Rover systems, and can honestly say it made a major difference in my career. Service on the Astronaut Rescue Team was a major factor in my selection for fire chief and other promotions as well.

The lesson here is to seek opportunities to increase your knowledge, develop skills, and accept those once-in-a-lifetime opportunities without succumbing to peer pressure.

Danny Krushinski

Go In and Get Her—Genuine Leadership
Before, During, and After a Line-of-Duty Death

Danny Krushinski has served as chief of the East Franklin Fire Department of Somerset, New Jersey, for 14 years. In his first several years, he set the example of how a chief should lead a department. A no-BS chief, he is very serious and focused about the mission of taking care of the public and his firefighters with training as his aggressive focus. He was doing it right—but then on April 11, 2006, Danny's and his firefighters' lives changed forever when he and his crew responded to a dwelling fire with a woman confirmed trapped. In minutes, crews went in to get her. There was a collapse and 21-year-old FF/Foreman Kevin Apuzzio and the woman he and his brothers were rescuing, both perished. This fire seemed to be "routine" in nature—a fire in a small single family dwelling with a victim who was easily reachable a short distance in from the front door.

Two lives were tragically lost mainly due to conditions that were nearly impossible to know due to the concealment of burned-through floor joists under a double layer of tongue-and-groove flooring. As noted by the expert investigators: Sometimes there are circumstances beyond the control of firefighters and the officers that command them that occasionally result in tragedies. I have been close to Danny well before the fire—and we remain close to this day. His amazing ability to move himself and his members forward is nothing less than spectacular. Danny has so much to "pass on."

Danny Krushinski

What You Don't Know Will Make the Difference between Life and Death

Go in and get her. Five words that changed a lot of lives. My name is Danny Krushinski. I have been a volunteer firefighter for 30 years. I have served as chief of East Franklin Fire Department since 2000, and also served as a chief officer in another department for five years.

On April 11, 2006, at 6:12 a.m., my fire district was toned out for a reported structure fire with entrapment. Prior to arriving, I requested a second alarm be dispatched and directly contacted the neighboring New Brunswick Fire Department to request their assistance. On my arrival, I found we had a 1½-story Cape Cod-style home with a moderate smoke condition. I found a police officer holding back the homeowner, an older gentleman, from trying to return to the inside of the house that was on fire. I was told by the homeowner that his wife was still in their bedroom, about 12 feet down the hall and on the right.

The first-due engine arrived on scene at 6:19 a.m., six minutes after dispatch with a crew of four fire fighters and one probie. The probationary firefighter from that rig pulled the preconnect 1¾ line, flaked it out, and left the nozzle at the front steps.

I quickly did a face-to-face with the crew of the first-due engine (Foreman Matt Desmond, Foreman Brandon Shannon, and FF Ryan Daughton) and explained where the victim was in the house. The handline was charged; the crew bled the hand line off and headed into the house to search for the victim. The crew of three that entered the house had their irons, TIC, flashlights, portable radios, and a short hook. The men were in full PPE with no exposed skin and SCBA. A water supply was being established for the first-due engine by the second-due engine, which arrived at 06:20—eight minutes after dispatch.

While they were laying in, the third-due engine arrived at 06:21—nine minutes after dispatch, with three firefighters and one probie. The three firefighters came up to me and asked for an assignment. I told Foreman Erik Wiklund to be the outside vent man (OVM) and for the two others, Foreman

Kevin Apuzzio and Firefighter Nick Recine to mask up and *go in and get her*. They made their way into the house with all of their tools, irons, TIC, and flashlights. The crew of the first engine located the victim in the back bedroom and started to try and remove her from the structure.

The crews went in and found the victim and were on the way out in less than seven minutes. While they were removing the victim from the bedroom, the crew from the second engine met them and began to assist with the rescue effort. As the crew neared the front door it was reported that the victim became hung up on something. It was at this point that there was a catastrophic failure of the wooden floor joists that burned undetected for an unknown amount of time, and the crew and victim began to fall into the basement.

Approximately 15 minutes after dispatch, as the interior crew was two feet from the front door and completing the rescue . . . the floor began to collapse. Foreman Matt Desmond attempted to jump over the collapse but only made it to the foot of the doorway and had to be pulled out of the hole by myself and a New Brunswick firefighter as Firefighter Recine dove out of the house from the front door. Foreman Shannon, Foreman Apuzzio, Firefighter Daughton, and the homeowner fell through into the basement, with furniture from the first floor falling on top on the men. A huge fireball came out and engulfed the entire first floor living room and hallway. A Mayday was transmitted for firemen down, and a fourth alarm was transmitted. The mutual aid truck from New Brunswick was on scene and tried to make its way in from the side door of the house while the rest of the crew was searching the second floor.

Foreman Wiklund ran back to the nearby first-due engine and grabbed a collapsible attic ladder, opened it, and threw it in the hole where the fire was coming from. Foreman Shannon emerged from the fireball off the ladder. He told me he dragged Firefighter Daughton, who was unconscious, to the bottom of the ladder. The firefighter assist and search team (FAST) went down and carried him out. A personal accountability record (PAR) was conducted, and we could not locate Foreman Apuzzio on the outside, although his helmet and rope bag were on the porch. A personal alert safety system (PASS) alarm was heard coming from the basement area. An intense rescue effort was put into place to get into the basement to get Foreman Apuzzio and the homeowner out of the building. The actual fire burned unchecked for 10 minutes while the remaining few firefighters who were on scene tried to get into the basement.

Foreman Apuzzio was located at 0702 hours—50 minutes after dispatch. Crews had great difficulty in accessing him due to his being trapped beneath the collapsed floor area in addition to high water levels in the basement from the prolonged hose line operations. They deployed portable pumps to dewater the basement and used several saws to cut through the collapsed debris. Upon reaching him, a rope was tied to him, and personnel pulled him from the basement. Foreman Apuzzio was removed from the structure at 0739 hours. This came 1 hour and 27 minutes after dispatch, and 37 minutes after initially locating him. Foreman Apuzzio was transported to Robert Wood Johnson Hospital where he succumbed to his injuries.

The initial crew of men that responded that ill-fated day on the first two engines were all under the age of 24, except myself and the driver of the first engine. These firefighters, all young and well trained but with very little actual interior experience, could be found almost every weekend attending some kind of hands-on training throughout the county, or just in the firehouse with the compartment doors open going over the tools and equipment and their uses. By these guys taking their training so seriously and wearing full PPE, it also prevented further tragedy.

Probationary Firefighter Josh Sullivan was taught how to properly stretch a line without any kinks. Not only did he pull the first line, he had the backup line ready to go as well. The first handline went into the basement with the crew during the collapse. The backup line held the fire at the door from the outside so Foreman Shannon could come up the ladder. Firefighter Wiklund and Foreman Apuzzio two weeks prior to April 11 attended a survival class and used a folding attic ladder in training scenarios. Foreman Shannon kept calm under extreme conditions and dragged Firefighter Daughton to the ladder. Although Firefighter Daughton never fully lost consciousness, he never removed his mask although he was almost fully out of air; he started skip breathing. Firefighter Daughton spent two days at a burn center for second-degree burns on his legs, neck, and back after he was removed from the basement. Foreman Desmond and Firefighter Recine received burns to their arms where the sleeve of the coat separated from the tops of their gloves. Foreman Desmond also had burns on his legs. His burns were received when he was hanging on by the ledge of the front door, dangling into the basement. Foreman Shannon miraculously received no injuries while being trapped in the basement.

While most 20-something-year-old kids in 2006 spent time on computers, X-Box, or PlayStation, these guys spent time training and making sure they could get the job done. We will never forget Foreman Kevin Apuzzio, for not just what he did on April 11, 2006, but how he was a big part of the crew's being trained so well and working with new firefighters so they knew what to do when they got off of the engine, that they always had a tool and were in full PPE.

I gave the order to *go in and get her*. I will have to live with the fact that Kevin lost his life under my command, but I know Kevin was the guy who had a lot of followers, and if it was not for all of the training he and the other crew members had together, we would've lost more than just Kevin that day.

Rick Lasky

Pride. Love. Passion. Recharging and Reminding Us Why We Became Firefighters

I'm assuming that most folks who buy this book have or had a love affair with being a firefighter. That great quote from the movie *Backdraft* summed it up well: "The funny thing about firemen, is night and day—they are always firemen." And that's true for most of us—we are firefighters 24/7/365. Rick Lasky is pretty passionate about being a firefighter. Actually, he's kinda nuts about it . . . and has been very successful in reminding us why—and reminding some that this may not be for everyone. I got to know Rick when he started instructing at FDIC, and it was evident that he loved this stuff. However, what really struck me was when I was invited to his fire department in Texas to speak at a memorial service. He actually ran a department that acted the way he described. *They* loved being firefighters—and the energy was phenomenal. While I'm sure they had some bad days there, the fact was that the gung-ho spirit was alive and well, and something of a great source of pride. Sometimes we "fall off the wagon" in remembering why we became firefighters—we lose that feeling we had when we were sworn in, graduated probie school or were promoted. Rick has assumed the role of "cheerleader" (without the outfit) in helping us regain that great feeling.

The Firefighter's Time Machine

Every firefighter with more than 30 years in the fire service (if they'll admit it), if given the opportunity, would go back in time and change a few things, maybe fine tune a couple of areas, whether it's how they walked into the firehouse that first day to just how hard they studied for that promotional exam, to trying to have a little more of an impact on those that worked for you and with you. Most firefighters would always want make it a little better. It's in their nature.

As much as they would like to do that, we know it's not physically possible. Or is it? You see we can accomplish it by passing on those lessons learned, those experiences, and yes at times the advice that some need to hear but don't want to. Don't you just wish you could make that difference? Have that chance to say "Hey, don't do what I did" or "Do this and you may just live to see your retirement!"

There are so many times I wish I could sit with that new firefighter and remind them that it is an honor and a privilege to serve as a firefighter—whether you're a volunteer or career firefighter, that it's more than what you see on TV or those tattoos and t-shirts. I know . . . very cool, yes . . . but a firefighter stands for so much more than just those things you see on the outside. It takes something on the inside as well—something that takes heart, devotion, a strong work ethic, and a life committed to values. Children look up to you and view you as a hero and more importantly as a mentor. If you don't believe me just take a ride to your local bookstore. Walk in to the children's section, and count how many books have a firefighter, fire engine, ambulance, or hook and ladder truck in them. Then you'll start to get it.

There are those times when I wish I could sit with those new company officers and remind them that it's about their people first, then them—that as an officer your firefighters come above everything else, and that good, firm, fair leadership is how we keep firefighters healthy and safe. As my best buddy Chief John Salka (FDNY ret.) would say, "You're not up there just to beep the horn!" It's about leading from the front. All those programs such as Courage to Be Safe and Everyone Goes Home do save lives, but it's even easier than that. You want it? Here you go:

- *Wear your seat belt!*
- *Slow down!*

- *Put your facepiece back on; this ain't your old man's fire anymore!*
- *Get someone to help you lift that patient before you blow your back out. Yes, I know; you'll have to wake up the engine company!*
- *Train, train, train!*

Seems kind of simple doesn't it? Well, it is. We have the ability as officers to keep our people alive. I know, bad stuff happens; but it's your job to reduce that amount of bad stuff. Lead by example and change your mindset from, "It's my job to make sure you go home at the end of the call or the end of the shift" to "It's my job to make sure you go home, at the end of your career, with a good back, two good knees, and cancer free." How about looking more long range down the road instead of just to tomorrow?

Rick Lasky

And yes, there are those times when I wish I could sit down with those chief officers and remind them that no one ever said being the leader was going to be easy—that it's hard making the right decisions because they're right and not popular, and that those right decisions don't come with cakes and pats on the back. Sometimes it's just the opposite, but in the long run you'll know what you did and the decision that you made was right because it was what was right! I'd remind them that it's lonely at the top and to make sure they had their "networking ring" set up. You know, those other chiefs you can call and confide in and ask if you're on the right track or off just a little. Last, be fair but firm, objective and open minded, and to remember it's about your people not you when it comes to the success of the organization.

In closing, if I had the whole group in front of me I'd remind them about what "it" truly is and that if more firefighters, officers, and chiefs truly understood what "it" was, a lot of things would just flat out get a lot easier. That "it" is less than those material things and more about how we got this whole fire service thing started in the first place, how we've gotten this far—our history, our heritage, what that Maltese cross actually stands for—"it." That "it" is the very result of what that word *service* is all about. They call it the fire service for a reason because it's about those people out there, those we put our hands up for and swore to take care of—them. Simply put, it's that "service above self" way of life.

So there you go. Love this job, take care of each other, and remember to leave it a little bit better for the next guy.

Edward Mann

A Burning Heart the Size of Pennsylvania

There are few states lucky enough to have a state fire commissioner like Ed Mann. First, he's an active fire officer—so when he speaks, he's current and timely. Ed is visible all over the state, taking care of the firefighters, fire chiefs, companies, and departments in whatever may be needed. What really drew me to Ed, and what has resulted in a solid friendship, is Ed's passion and energy to reduce unnecessary firefighter line-of-duty deaths and injuries. Several years ago, Pennsylvania was the highest in numbers related to firefighter LODD—and that jolted Ed. Knowing so many of the deaths and injuries were avoidable, and with him personally attending nearly all of the funerals, action was needed. Working with numerous Pennsylvania career and volunteer organizations, he has been successful in leading the Pennsylvania fire service through numerous educational programs focused on firefighter safety, health, and survival. If each state had an Ed Mann, we would have much less bad stuff to write about.

Edward Mann

Simple Truths, Hard Lessons

Be honest and don't lie.

You may decide not to say everything, but don't ever lie to the troops—in the end, the truth will surface. You will lose your credibility as a leader, and it will take years to regain their trust. Sometimes being honest can be an unpleasant and painful experience, yet it has to be done. I also believe it can be done in a tactful way so the other person still has some dignity left when you're done. Case in point: Over the last several years I had to tell several fathers that their sons were not ready for the fire service. While teaching entry-level fire training programs it became obvious to me that the young men were in the class because Dad wanted them to be. In the end they failed or dropped out from the classes, and their dads wanted to know what the issues were. It was difficult to tell fellow firefighters—guys I had known for several years—that their sons had no interest in our business at this point in their lives and should really find another calling. In the end the young men were better off for it, and the fire service was better served.

It is okay to say I don't know.

As a firefighter, company officer, or chief officer, there will be times you simply don't know the answer to a question. It's okay to say, "I don't know." If you attempt to make up something just to save your own pride, you will eventually be exposed as a fraud.

Don't try to impress me with what you know; teach me what you know.

Regardless of the position you have in the fire service, you will no doubt be asked to teach others. Don't squander an opportunity by trying to impress students with what you know—teach them what you know! I recall early in my career, I attended pump training and, of course, hydraulics was a large part of the class. One instructor strutted around, spouting off formulas and throwing out numbers like a professor. Well, the only thing I learned was I did not care for the instructor. Sure, he was knowledgeable, but he taught me nothing about hydraulics. On the other hand, the other instructor not only taught us what he knew, he inspired the class to want to know more. I found myself going home at night and practicing the methods he conveyed that day in class.

It's okay to say no.

I grew up in the fire service being taught that an aggressive interior attack is the only way to put a fire out. I'm now of the belief that it's okay to say "no!" I'm not putting my people into a lightweight constructed home that is being attacked by fire, especially when there is no one to save. Obviously, it's not that simple—it requires an understanding of building construction, where the fire is and where it's headed, fire flow versus water available, and most importantly, if there's human life that can be saved.

I would rather defend my call not to send my troops in than stand in front of a heartbroken family because one of my firefighters died when the lightweight home collapsed. For those of you who disagree, all I can say is this: do as you wish on your fireground, but do as I say on mine.

Don't let something bad wipe out all the good you have accomplished.

Over my fire service career, I have had the opportunity to work with some really terrific people ranging from firefighters to fire chiefs, and many of them had stellar careers in paid or volunteer departments. Unfortunately, I have seen some of them do something stupid near the end of their careers that wiped out all the good they had done. For example, some chiefs I have known made their departments into well-respected, well-trained, and well-equipped organizations; they always seemed to be able to convince government officials to support their cause and were admired and respected by the citizens of the community they served, only to have it all come undone because they became too set in their ways and refused to keep pace with the changes taking place around them. In the end they either left office or got voted out, and nobody remembered the good they had done.

Surround yourself with honest people.

You should surround yourself with people who are willing to challenge when needed, people who are willing to tell you why they believe you're about to make a mistake or why you should go another direction. In the end, I believe you're better off for it.

Be happy, have fun, and find something else to do when it isn't.

Whether you're a volunteer or paid to do what you do, if you're not happy doing it or having fun doing it, find something else to do. You'll be better off, and the people around you will be much happier, too.

If the organization will be better off, and the people you serve will be better served, then the answer has to be yes.

Many times over my career, whether I was serving as a manager or voting on an issue at a company meeting, I had to make decisions. I learned that if I answer yes to these questions, I have to vote yes, regardless of my personal feelings. In the end it's not about you; it's about the good of the organization and the people we serve.

Don't hold grudges.

I have seen grudges among family members, firefighters, and entire organizations ruin the same. Holding grudges makes for good western movies but has no place in our business.

Be safe for your families' sake.

No explanation needed.

RICHARD MARINUCCI

From the Classroom to the Fireground

Rich and I have been friends through what seems like several lifetimes. I mentioned in this book about the late Don Manno. Rich was part of that close circle of friends and was there with us when Donny was laid to rest. Rich was a schoolteacher before he became a fire chief, which may be some of the best training and qualifications to be a chief. His high energy and enthusiasm are well known. He currently serves as chief of department in Northville Township, Michigan, and before that he served nearly 25 years as the fire chief of Farmington Hills, Michigan. A past president of the IAFC, he also served as the acting U.S. fire administrator while serving as senior advisor to the FEMA director. Are you familiar with "America Burning?" Rich provided testimony during that period to the congressional committees and supported the creation of the Blue Ribbon Panel, America Burning Recommissioned. He was one of the first 15 chief fire officers to be designated as "CFO" in 2000 when the accreditation process started. From the classroom to the fireground to the federal level, Rich has so much to pass on to us.

Linda and Rich with Reagan and Sam. Not pictured are Rich's three other children: Jeff, Jill, and Jessica.

What I Have Learned—Thoughts and Advice

Reflecting back on a lengthy career in the fire service and identifying a couple of specific items to discuss relative to "what I have learned" may seem a bit daunting. Sometimes a review like this actually reveals a few key things that surface over and over again. In this case, it's education and training along with relationships that are the main themes of my brief discussion on the topic. Attention to these subjects leads to success in the fire service and to personal development that transcends the fire service.

Offering education and training as an important component of success does not qualify as earth-shattering advice. "Duh!" would be a typical response. But if it is so obvious, why doesn't everyone pursue it? Too often individuals attain their goal and lose sight of the long run or the big picture. People can become complacent or content with their lot in life and remain in their comfort zone. For those who want to pursue a career that offers tremendous gratification, education and training must be a significant part of the entire process until the individual retires from service. The future is unpredictable, but preparation for the future allows people to be ready when opportunity strikes.

Education throughout life is critical. In this context, education is different from training. It is intended to broaden one's perspective, improve critical thinking skills, and expand the mind. All of this keeps an individual sharp and better prepared for whatever life may bring. Although there are no guarantees, educated individuals have a much better chance of success. As much as someone plans for the future, there will always be changes to make. The world is always changing, and opportunities come and go. Those who are best able to adapt to new environments have the best chance to succeed. Individuals must not postpone their education, as the longer one remains out of this learning environment, the more difficult it is to return. Of course, balancing all of the parts of your life is extremely important. Be consistent and moderate in all aspects of your life, both personal and professional.

The fire service is as dynamic now as it has ever been. This means that it is changing more quickly and dramatically than ever before. Training is not a luxury and is absolutely essential in developing and maintaining competence in the ever-growing responsibilities assigned to fire departments. Training is applicable to skills. Basic firefighting training starts the process but is only the beginning. It is imperative to maintain competence commensurate with the expectations of your organization and your own personal desires. It also involves the repetition necessary to ensure that the best possible service is delivered every single time. You must train in the basics and also for the occasions you will rarely face. This is extremely important for the public being served and also for you, as it is how you will succeed and also how you give yourself the best chance to minimize risks in a very risky profession. Continual pursuit of excellence with the goal of perfection should be every firefighter's objective throughout his or her career.

Successful people, regardless of the profession, understand the value of relationships. Establishing a wide-ranging network internal and external to the fire service greatly enhances an individual's abilities and improves capabilities in "getting the job done." This is most easily accomplished by remembering the things you learned from your mom and while in kindergarten. Be nice, be polite, say please and thank you, control your ego and emotions, treat others as they would like to be treated (usually how you would like to be treated), admit your mistakes, apologize when necessary, be a professional so you don't make too many mistakes, be on time, make your deadlines, don't make promises you can't keep, and do what you say you will do. This list should only be a start. Compile your own as a way to enhance your ability to build relationships.

Over the years I have met an unbelievable number of really smart people in the fire service. Even with special talents, the overwhelming majority of those who are successful have built relationships throughout their careers. Few could get by just on their own without the support of many others. I would also say that I have met folks who I considered having average talent but were very good at building relationships. They have enjoyed a great deal of success by understanding that interpersonal dynamics are extremely important when pursing individual and organizational goals.

As firefighters ascend through the ranks, they have more opportunities to interact not only with those in their organization but those outside of it. This would include those inside the fire service as well as those who do not have a direct connection. So much of what gets done in this world is based upon relationships, even more so than logic and/or sound arguments for or against. Some people can compile the greatest discussion points but are unable to convince others of their viewpoint, not because they are wrong but because they did not understand the value of relationships.

Remember, you are always making an impression on someone else. You must learn to put your best foot forward all the time. It is who you are. You must be professional. By this I mean you must know your profession as well as possible so that you can clearly be identified as an expert. You must treat everyone you meet with the utmost respect. You must take the high road when dealing with issues and "do the right thing" to maintain your integrity.

Those who I've met throughout my career, who I consider the most successful, are very knowledgeable and possess great interpersonal skills. These attributes are within your control. You can educate yourself, and you can be nice to everyone. Not only have these individuals been successful, they had a profound influence on my career. The lessons mentioned here have been learned from others, some through discussion and others through observation. Keep your eyes and ears open, and learn from those who you deem worthy.

Ray McCormack

Keeping Fire in Our Lives

I don't hang around people who don't love this job—and I avoid those who aren't passionate about it. I depart from the downers and run from the whiners for the same reason. Go whine to someone else; I am not interested. What I do try to do is hang around people who are passionate about this stuff, whether or not I totally agree with them. Like nearly every contributor in this book, I agree with a lot and don't always agree with everything. I got to know Ray over the last several years when he started teaching at fire conferences—he is good to hang around with. Ray has forced us to focus on a "culture of extinguishment." You know, getting water on the fire. I challenge anyone to argue with that logic. I would add that his focus includes a culture of training, leadership, and constantly being a student of this job. It's never-ending adult education. Ray's culture of extinguishment means well-trained firefighters and officers focusing on saving the lives of those who call us—while doing our absolute best to protect our own—as the mission of the fire service. When we lose focus on that, we might as well go home.

Ray McCormack

A Culture of Extinguishment

The role of a fire department is suppression and extinguishment. It is why we exist. If you disagree with that statement, then turn in your fire trucks. An overly dedicated all-hazard mindset, along with other distracting mandates, can work to erode a fire department's culture of extinguishment. The disconnect from embracing a robust culture of extinguishment begins with a perceived notion that extinguishment must be accomplished at all costs. That, of course, is false. There is barely any definition for a culture of extinguishment, never mind a standard. That being the case, let's define and explain what it means to most of us and what it is not. It is about commitment to the art of suppression and perseverance during suppression, two qualities firefighters need to employ to fulfill the department's mission to the community and the fire service itself. It is about the constant discovery and refinement of the suppression mission. It is about working to provide the best in suppression tactics and efficiency. It is about saving lives, both civilian and firefighters. It is about property conservation. It is not about some twisted sense-of-duty, risk-it-all mantra, which many have tried to ascribe to it.

Extinguishment is accomplished with different approaches, tools, and line sizes, and from both inside and outside the fire building. Where it has gotten messy is when people attach firefighter risk and injury to a culture of extinguishment, the premise being that firefighters will risk it all for a chance to be close to the action. While that makes for a clever narrative and may happen on occasion, it is only a cultural fault because of a training deficit. Training, command, and control are part of the checks and balances that place firefighters in correct positions. If we look at root injury causes correlated with extinguishment, you often find an underdeveloped extinguishment culture.

A culture of extinguishment is something we must have in the fire service. To claim that a fire department's mission to preserve life and property from the ravages of fire should not foster a culture of extinguishment is farcical. Remember the culture of extinguishment belongs to the fire service. We created it.

Do fire departments do more than just fight fires? They surely do. Are there operational concerns other than extinguishment? There surely are. Fire departments are all-hazard these days, but remember that despite name changes and high medical response rates, only firefighters put out a community's fires. If the fire service does not support a robust culture of extinguishment, then our communities are being shortchanged.

The fire service is the most valued public service entity there is, and a strong culture of extinguishment keeps it that way. If you do not like how some perform fire attack, then work to change it for better. Focus on apparatus design, attack standards, hose loads, and what can be done to stay focused on improvement. If you are a firefighter and believe that a culture of extinguishment is not important to you, your fellow firefighters, your department, and community, you have somehow been led astray. If you are not reminded by seeing fire engines in your station or wearing your gear as you respond that we still put out fires, take a break and refocus. Suppression is why we exist, no matter your call volume.

The culture of extinguishment defines effective fire departments. Our service to the public comes in many forms. Suppression, when called upon, must be done well. If you see a fire department that does not perform well, most likely it has yet to fully embrace a culture of extinguishment.

Keep fire in your life.

Bruce J. Moeller, PhD

Firefighting and City Management: Oil and Water or Valuable Opportunities

I've never kept a very low profile regarding most of my relations with city managers. Sadly, in spite of my well-documented diplomatic attitude, sometimes it's been a challenging relationship. However, I have had some excellent bosses over the years. There are city managers who understand and appreciate the positive value of a well-led fire department. My old friend Bruce Moeller is one of them. Bruce and I actually both worked in the same city—but decades apart. Bruce is a progressive thinker who went from firefighter to officer to chief—and then, he stepped across the tracks and became a city manager, but was still a firefighter at heart. This is very rare. Based upon my past relations with some city managers and my long-term friendship with Bruce, any fire chief would be lucky to have Bruce as a boss.

Camp Washington Neighborhood Firehouse, Cincinnati, OH

Think!

My grandfather was in the fire service. I never knew him personally, for reasons of generations and geography. Later when I found this photo of him at the opening of Cincinnati's Station #43 in 1908, I was strangely both connected to and disconnected from my family's fire service history. That is how the fire service has always been for me—connected and disconnected. My advice for you is this: You should try to make it the same for you.

The connectedness is from our shared history and brotherhood. It is steeped in shared experiences and traditions; and what we do today is often based heavily, if not exclusively, on what was done in years gone by. When we first join the fire service, we are indoctrinated into these rituals and find comfort as a young baby finds comfort in the embrace of its mother.

For me, the disconnect arises from unanswered questions, what I have come to call Big Questions for the fire service, questions about why we do what doesn't seem to make sense. If response times are so important, why do we slow ourselves down—especially during the night? Why do we take risks trying to save buildings that are empty and already lost? Why do we continue to work the same way we did decades ago?

I'll assume you understand the *connected* part of this essay. We all want to be recognized, part of a unique group, honored for our work, and so on. The fire service provides this, often in a special way that extends beyond the firefighters themselves, but also to family and friends. How many times are you aware of those around you boasting on what we do? It is one of the great parts of being a firefighter.

So let's talk about being *disconnected* from the fire service. What does that mean? Well, if you really love the fire service, you must—at times—pull back from it. Disconnect!

As a member of the public safety services, we are indoctrinated to obey the commands of our superiors. We are taught to listen to the directions and comments of those with greater seniority. The problem is, what happens if they are wrong? On the fireground we can't have rank-and-file members questioning the incident commander. However, off the fireground, we should take a moment to question if the information we are being told is credible—does it make sense. In other words, you need to temporarily disconnect from your responsibility to follow the leader and apply some *critical thinking*. I have a brother who learned how to accomplish this when he was a bartender while in college. Sometimes he was dealing with the establishment's resident philosopher; more often, it was the local know-it-all who would offer all in attendance some type of outlandish claim. Regardless of which type of statement was made, he simply replied, "Is that so . . .?" Not agreement or disagreement, simply an acknowledgement of the statement, and meant to provide the opportunity to further consider the potential value of the comment or perspective offered.

This type of I-heard-you-but-let-me-think-about-it-for-a-while response avoids what Irving Janis identified in 1977 as *groupthink*, where the need for conformity results in a deviant or perverse decision outcome. In the fire service we are especially prone to this social disorder, in part because we are often xenophobic—afraid of any thought or idea that come from any source other than the four walls of our local fire station.

Take time to challenge the conventional wisdom. Not in a confrontational manner, but in a show-me-or-prove-it approach. In short, take time to think. If you love the fire service, we need more of that.

Frank Montagna

Four-Plus *Decades* of Responding to Routine *and* *Not-So-Routine* Fires and Emergencies.

You have probably read *Responding to Routine Emergencies* by Chief Frank Montagna of FDNY. If not, consider that your homework assignment because Frank digs into the so-called "routine" runs that many of us have had little training on. I became friends with Frank when we taught together along with Jim Murtagh at FDIC. What struck me most about Frank was his calm demeanor and humor, a quality that I consider pretty important. With 43 years on the FDNY, he served as a company officer since 1979 and as a battalion chief since 1987, commanding the busy 58 Battalion. He finished his last few years (prior to retirement this year) assigned to the FDNY Bureau of Training, where he developed training courses. He developed and taught the department's new Battalion Firefighter Training Course, created the Captain's Development Course, and has introduced simulation training to many of FDNY's courses.

Frank Montagna

Training Yourself

There is so much advice that should be given to new firefighters it is hard to know where to start and difficult to choose what to suggest in a short article. I want to give you advice that will serve you well throughout your career, not just help you out of one tight fix. My advice needs a bit of explanation, so here it goes.

There are only so many scheduled training hours in our day. That is true for both paid and volunteer firefighters. It is also true for large departments as well as small ones. The scheduled training that we receive sometimes is great, but sometimes not so much, and some of those training topics are mandated by various federal or state agencies, not chosen by our own department based on our own needs.

Over the years, I have repeatedly encountered situations for which my training hadn't prepared me. These—let us call them *surprise incidents*—were sometimes life threatening and sometimes not. They included structural fires, utility and other emergencies, hazardous materials incidents, unknown odors, outdoor rubbish fires, car fires, leadership/management problems in the firehouse, and more.

Some of them had to be acted upon immediately. Some of them allowed me time to consider and consult with others before taking action. Most left me with unanswered questions. Often those questions remained unanswered by my department. Don't get me wrong—my department has a great training program. It has improved greatly over my 43 years of service, but all training programs have failings, even yours. So what can you do?

Here is my advice, and you might not like it. It requires you to do work, and not just a little.

When you encounter something that you don't know the answer to, or don't understand, find the answer. Ask questions; do research. Don't quit until you have the answers you want. That is it. It is simple, and it is hard. The obvious questions you should be asking in response to my challenge are how and where to start. In answering your questions, I'll tell you what I did.

First, ask questions. Ask your officers and senior firefighters. Ask firefighters from other departments. Ask the utility workers. Ask building management and maintenance people. Whatever the incident, find someone who deals with the subject regularly, and ask them what you don't know or understand. But don't accept their word as gospel. You have to dig deeper to be sure you have gotten all the needed facts. Go to the Internet and search the topic. For most everything you want to know, you usually find a wealth of informative sources there.

Another excellent source is the books written by firefighters, for firefighters. Visit the websites http://www.Amazon.com or http://www.pennwell.com/index/fire-engineering.html, where you can find books that may answer your questions. Again on the Internet, look for firefighter forums and training communities like http://www.fireengineering.com/index.html where ideas are exchanged and questions are asked and answered by firefighters from around the world. There is also a number of excellent firefighter-related publications you can subscribe to. They supply a continual stream of firefighting knowledge to anyone who is willing to read. Still another way to get the knowledge that you need to stay safe in our dangerous job is by attending the various training conferences offered around the country. If you only attend one, I suggest the Fire Department Instructors Conference held annually in Indianapolis, Indiana.

What I am suggesting that you do will cut into your leisure time. It may cost you money. It will require that you do work and even that you read. It can, however, provide you with a wealth of knowledge and possibly save your life.

Oh, I have one more thing to add. After you get the answers you are looking for, spread them around. If it was something that you did not know, then you can bet others were in the dark about it. Tell them. Better still, pick up a pen and write an article about your research, and tell the whole darn fire service what you learned.

Tom Mulcrone

A Father in the Fire Service

I am friends with many fire department chaplains of numerous faiths, and they all understand the need to take care of their "fire flock." A trained chaplain can make a positive difference to a department by knowing how to ride the wave in an organization, tending to all of its needs, as a group or one at a time. A good chaplain understands that each member has different needs when things are tough and individually serves the needs of a member. Father Tom Mulcrone is the well-known and beloved chaplain of the Chicago Fire Department, serving the firefighters and paramedics for a few decades. As I am not a Chicago firefighter (although I was born in the old Cook County Hospital), I have never experienced the services of Father Tom—but I know those who have. "What they're doing is God's work," he says. "They always say, 'I was just doing my job,' so I remind them that what they do is sacred."

Tom Mulcrone

A Sacred Task

I think it is safe to say that almost everyone in the fire service remembers their firsts—first assignment, first day on the job, first officer, first run, first (and usually only) nickname. The people and events that firefighters and EMS personnel first experience leave lifelong impressions.

While a part of a department, the chaplain's role is quite different from that of sworn members, so our firsts are memorable in significantly different ways—the first chief who brought you on board, the first firehouse visit, the first religious ceremony performed as the new chaplain, the first firefighter or medic who came to you for help, and, yes, we are not immune from that first and only nickname. And for those of us charged with the spiritual care of a department, it is the first line-of-duty death that is never forgotten.

For me, it was a young firefighter, not yet 30, who was fighting a church fire on a bitter cold winter night in Chicago. While trying to rescue a caretaker supposedly still in the building, there was a smoke explosion and catastrophic collapse of the structure. The rest of his crew just barely made it out, suffering injuries themselves. When he didn't, the grim job of digging through the rubble began; and while we all hoped against hope, the magnitude of the collapse forced all of us to face an awful reality that night.

With news cameras all around about to go live, I met with the commissioner and told him that I was going to notify the family. Then, accompanied by a chief, I stood at the front door of the family home and, filled with dread, rang the doorbell.

For the next few days the Chicago Fire Department was busy with planning a full-honors funeral, and I had to care for our fallen brother's family as well. Because it was my "first," there was a significant learning curve involved. And even though I prayed there would never be another, it was then and there that I pledged to myself that caring for the survivors of the fallen would be a priority in my ministry as chaplain.

Some months later I ran into a widow of an LODD firefighter—her husband had died along with two others from his company two years earlier. She spoke with great pride of her husband, of the service he rendered, and of his ultimate sacrifice. But it was with sadness that she told me about no longer feeling a part of that "second family" of which she had always been so proud. It seemed that after the funeral was over, so was her relationship with the department she loved; truly, it pained her.

When any of our own falls in the line of duty, we go out of our way to honor them and care for their families at the time of the funeral. We use phrases like: "We will always remember" and "We shall never forget," but sometimes, through no one's fault, the bonds we forged disappear and people are forgotten.

It was because of this that we were blessed to establish the Chicago Fire Department Gold Badge Society as a support group for the families of our fallen brothers and sisters. Through their advocacy for duty death families and their presence at department events, these surviving family members have become an integral part of our department's family.

I can think of no more rewarding work of a fire service chaplain than caring for the families of our line-of-duty-death members—truly never forgetting our fallen brother or sister, or the ones they have left to the care of their second family.

Jim Murtagh

Decades of Fires . . . and Learning Every Step of the Way

In 1995 I reached out to a friend I deeply respect, retired FDNY DC Sal Sansone, to help provide some training to our department officers in Ohio. He brought with him Chief Jim Murtagh—and our friendship began. Jim has since retired as FDNY's assistant chief of department, serving previously as commander of The Bronx, Brooklyn and the Bureau of Training. Actually, there are few positions that Jim didn't hold within the FDNY over his decades of time on the job. While personable and always with a joke, he commands a kind of respect that reminds me of my Dad. There is a time for fun, but there is a time for business, and Jim is all business when it needs to be. He, like others in this book, is truly a student of this job, and even as a retiree he continues learning. Last year, for example, while preparing for a course he was delivering, he reached out to Steve Kerber (another contributor to this book) and the crew at UL and NIST so he could learn for himself about their latest studies on fire behavior. That stood out as impressive to me—he's never stopped wanting to learn. Jim is a graduate professor and instructor at numerous universities and colleges as well as the National Fire Academy, and he taught for many years at FDIC. As the team leader of FDIC Instructors, he taught such classes as "The First 10 Minutes and Beyond," a 16-hour course for company officers, and "Boot Camp for Chief Officers."

The point is, he sets the example of how we must never stop learning as firefighters and officers. If Jim taught me anything, it is that the day any of us stops learning on this job is the day we should get out.

What Can You Do?

The fire service has always dealt with change, but from time to time events occur that create worldwide change: the Great Fire of London, the Great Chicago Fire, the Baltimore Conflagration, the Triangle Shirtwaist Fire, and most recently the 9/11 attack on America.

What can *you* do to prevent these events? The scope of that question is far beyond the scope of this essay, but the what-can-you-do part of the question is not.

The following are three important things to consider when preparing yourself for the next event that will be coming.

HAVE A FIRE SERVICE LIFE PLAN

Early in your career decide what your ultimate goal will be. This is the "what do you want to be when you grow up?" decision. Not everyone wants to be a chief, nor does everyone want to remain a firefighter or become a company officer, but some do.

Reflect on your personal aspirations, then admit these to yourself and commit to making a Fire Service Life Plan that meets your needs. If you choose to remain a firefighter, then become technically competent at all firefighter roles, and specialize in those areas that you want to be the best at. This will be your Life Plan. It will take some time to achieve it.

If you choose to become a fire officer, you are taking a different route; you must be proficient at all firefighting functions, fireground operations, and the roles of company and chief offices. You will need to expand your learning into areas beyond firefighting skills and techniques; this requires extra effort and strong dedication.

If you aspire to be the chief, your outlook and preparation is significantly different still. While you still have to understand the what, how, and why of the roles and activities of firefighters and company officers, you need not become an expert in performing these tasks. You must have a clear working knowledge and understanding of their roles and how they relate to an effective fire service delivery system. At this level the fire service system becomes more prominent in your thought processes. You will need a working knowledge of *all* the aspects of leadership, firefighting, and fire command. To be an effective chief officer you have to go beyond the company-officer levels of preparation and embrace the concept of going outside the fire service to gain knowledge and perspective. This includes the understanding of management, personnel, fire science, fire protection, fire engineering, and the emergency management fields.

All of these three potential Fire Service Life Plans have many things in common, but they also have significantly different Life Plan perspectives. Whatever Life Plan perspective you choose, it is an important one. It's also important to know that you don't have to choose what others have chosen. Each Fire Service Life Plan is different, but different doesn't make them more or less important, only different. It's your choice, and you will be respected for it. If at some future time your Fire Service Life Plan needs to be adjusted, then it's your choice, do it. The fire service needs people at all levels, people who are there because that's where they want to be, and they are confident in what they do.

INTERNALIZE THE FIRE SERVICE AS A LARGE TEAM COMPOSED OF MANY SUB-TEAMS AND YOU AS A TEAM PLAYER

When you entered the fire service you were probably told that you were part of a team, and you were proud of that moment. However, you were probably not told how big the team is, and of the various perspectives of the team players and leaders. The fire service has always been composed of team-oriented organizations; however, much has changed in recent history, and the team concept has grown exponentially, with numerous new and challenging roles. We've moved from being a local service provider to a large community service provider and even a national fire service provider. We've graduated from a local individualized response teams to preplanned fire service delivery teams, using practiced standard operating procedures (SOPs), incident command systems (ICS), and a National Incident Management System (NIMS). We've moved from being a small group that determines its own ways and methods of doing things to a community-wide mutual aid team. We've adopted common goals, common practices, and mutual agreements that outline how we will work together as a team. Our teams range from hose teams, ventilation teams, search and rescue teams, and emergency management services (EMS) to division, sector, branch, and command teams. We have and are being included with other emergency teams such federal command support teams, law enforcement teams, health teams, and hazardous materials teams. As a result of our nation's disaster on 9/11, we learned that our teams are no longer local, and our interaction with other fire service teams is not competitive; it's cooperative and supportive. A good example is our national urban search and rescue teams. They are good as a single team, but even better as mutual support teams.

DO ALL YOU CAN TO IMPROVE YOURSELF, BUT ALSO MAKE A STRONG EFFORT TO IMPROVE THE MEMBERS OF YOUR FIREFIGHTING TEAM

To achieve our goals we need to do all we can to improve ourselves and our fellow team members. The theory is: You need your team members to be at least as good as you are, and probably better; they are the people who will come to your aid when you are in critical danger. I call this the *selfish model*, because it stresses that your actions of training and preparing your fellow firefighters is for your own benefit. It's true that many others will benefit from your training and preparing other firefighters, and maybe you will never need them, but if you do, you want only the very best. I believe the ultimate goal is for you to create a strong safety backup plan for when things go wrong.

When you became a team member you, consciously or unconsciously, subordinated some of your personal interests and feelings and made a commitment to become a team player. You accepted the good-for-the-team concept, and you agreed to work with and for the team. Helping your team members get better is half of the task of this recommendation. Getting to know and being familiar with the members of other teams, other fire departments, and fire service social and formal organizations, as well as other emergency groups and organizations is the second half of this recommendation. The more team members you are familiar with, the better the team output will be.

Chief Alan Brunacini, when referring to firefighters' interaction with other firefighters and emergency service organizations, likes to say, "We need to learn to play well together in the sandbox." Knowing people ahead of time makes it easier to work with them; having worked or trained with them makes it even easier and better.

While I strongly believe the three activities I've written about are important to our mutual success, we need to recognize that they take time, a great deal of effort, and they may not always turn out as we planned or hoped for. However, with teamwork, team support, a personal Fire Service Life Plan, and the willingness to work and contribute to our fire service systems, we will overcome the hurdles in our path. The fire service is a great vocation, and it will continue to be our nation's first line of response when danger strikes. Be prepared, be safe, and enjoy your calling.

Jim Murtagh

Gregory G. Noll

Firefighter? Sure—But He Knows All the Other Stuff Very, Very Well, Too

I have always been impressed with people who strive to learn every possible aspect of the job. Greg Noll, a third-generation firefighter, is one of those people. Greg has served as a subject matter expert for various DoD hazardous materials and counterterrorism response training programs. He's coauthored nine textbooks and has received many awards, including the Chief John M. Eversole Lifetime Achievement Award for his national leadership in hazmat operations. Have you ever heard of the Yvorra Leadership Development (YLD) Foundation scholarship program for Fire and EMS personnel? Jim Yvorra was a deputy chief with Berwyn Heights (Prince Georges County, Maryland) and a nationally known author and editor. He died in the line of duty when he was struck by a car while operating at an accident scene on I-95 in January 1988. Greg led the way to honor his friend Jim and formed the YLD Foundation—giving away over $140,000 (www.yld.org). That alone says a lot about Greg.

Greg and family at the commissioning of 2nd Lt. Sean P. Noll (USMC)—Virginia Military Institute class of 2013

What I've Learned

I've had the two best careers possible. First, I've had the honor and privilege of wearing our nation's uniform, ultimately retiring with 29 years of service with the U.S. Air Force—Fire and Emergency Services. Second, I've served in both the fire service and the emergency response community for the past 43 years. Both of these careers have shared one trait—the honor and the willingness to serve. From where I sit today, there is no higher calling. We are part of an elite group that makes up less than 2% of the American population.

I've learned that a person's career is shaped by many individuals—parents, family, supervisors, peers, mentors, and others. Sometimes we don't recognize that we are sitting in the classroom of life until many years later. I've been fortunate to work with and for many individuals who gave me the opportunity to run with the ball, screw up and make mistakes, and then learn from those mistakes. It's easy to be there when things are going well in life, but life's lessons are usually found when the challenges are greatest, when perhaps only you know the depths of your soul.

One of my classroom-of-life teachers was Major Jack Jarboe of the Prince George's County, Maryland, Fire Department. Jack was a tremendous person to work for—one always knew where you stood in the game. But what I remember most about Jack was his simple saying about life's priorities: "Remember the 4 Fs—family, faith, friends, and the fire department . . . and in that order." I've tried to follow the 4 Fs . . . I can't say I've always been successful, but as I enter the final quarter of my professional career, I think I've gotten it right more often than not.

The military and the emergency response communities bring together some of the finest people you will ever meet during your career. Whether you call it the brotherhood or the sisterhood, good things happen when good people come together. It might sound corny, but it's true.

My classroom-of-life graduate course on character was taught by my good friend, the late Chief John Eversole of the Chicago Fire Department. When John passed away in 2007, many commented that "he never really acted as important as he was."

John Eversole led by example. A 19th century evangelist said, "Character is what you are in the dark." The most difficult choices in life, between honor and dishonor, are when no one is watching, when only you know if you have done right or wrong. Throughout our lives, God grants us all the privilege of having our character tested and our honor affirmed. The tests come frequently. We all fail some, but hopefully not most of them.

John Eversole exemplified character. As a peer, I believe the things that John was able to accomplish both personally and professionally emanated from the core principles that his parents taught him as a young man. Call them the basics of life—honesty, integrity, commitment, humility, and compassion. The highlight of my professional career was receiving the 2011 John M. Eversole Lifetime Achievement award from the International Association of Fire Chiefs—Hazardous Materials Committee for leadership and contributions to further and enhance the hazardous materials emergency response profession. The award is the Cooperstown of the hazmat response community.

My final graduate course in the classroom of life was family and relationships. I've had a very successful professional career in both the fire service and the military. But as a senior citizen at the age of 60, these are the memories I treasure most:

- Celebrating 25 years of marriage with my wife and best friend Debbie (aka The General).
- Walking my daughter Kendra down the aisle.
- Having my oldest son Sean serve as my retirement official for my USAF military retirement ceremony.
- Watching Sean commission as a second lieutenant into the U.S. Marine Corps, and then being his first salute.
- Having each of my children (Kendra, Sean, and Ian) tell me at different times of their lives that I was their hero.

In closing, enjoy your professional career as a firefighter, but don't forget your family. It may not seem like it today, but the ride will be a quick one. Enjoy it!

DENIS ONIEAL

From a "Jersey Boy" Street Firefighter to The Nation's Chief Fire Instructor—Setting the Example Every Step of the Way

Some firefighters who have never met him may not realize it, but the man who leads the nation's fire training has crawled halls, pulled ceilings, stretched lines, and commanded thousands of working fires. He spent 15 years in busy ladder companies as a firefighter, lieutenant, and captain and then went through all the command ranks. He also sets the example related to training and education. He walks the walk and talks the talk—with decades of "going to fire" to back it up.

I first met Denis in the 70s when he was a lieutenant in Jersey City, and I was working on becoming one across the river on Long Island. He ran a side business called FireEdcon Books with a friend, Lieutenant Jim Cline, E-10, FDNY. We've remained friends since the 70s and have been blessed to teach together all the way from Maryland to Hawaii—and many places in between. One of the smartest people I know is also one of the most "stand-up" and loyal gentlemen in our business. Denis is a solid friend, husband, father, and grandfather who has never forgotten his priorities or his roots as a firefighter.

Dr. Denis Onieal has served as the NFA superintendent in Emmitsburg since 1995. He began his career as a firefighter in 1971 for the Jersey City (NJ) Fire Department, rising through the ranks to become deputy chief in 1991, and acting chief of more than 600 firefighters. He has spent his entire career "in the street" as a line fire officer. He earned a doctorate of education from New York University, a master's degree in public administration from Fairleigh Dickinson University, and a bachelor's degree in fire administration from Jersey City State College. He was professor in the master and doctorate degree programs in education at New York University prior to his appointment. He has authored more than 30 fire service publications and, using the "checklist" format, has some great tips to pass on to you.

A Range of Thoughts from a Very Diverse Career

FIREFIGHTING

- Put your gloves in your pockets. Do your size-up while putting them on slowly. Don't make a decision or say a word on the radio until your gloves are on.
- If you can't control yourself, you can't control the incident or anyone at it.
- Control the stairwell; you control the building.
- Call for help before you commit your last company. If you don't need the help, send them home. They won't mind.
- Remember, when there's a crisis (Mayday/serious injury/LODD), you're still responsible for the incident! You have to assign someone to either take command of the incident or take charge of the crisis—you can't do both.
- When you're responsible for an incident, keep asking yourself these three questions: What's my problem now? What might happen next? What do I need to do about it?

MANAGEMENT/LEADERSHIP

- Rank is a responsibility, not a privilege. You're not promoted to "get" something from the organization; you're promoted to "do" something for the organization and its people.
- Rank is the last refuge of a scoundrel. When you have to point to your collar, you've already lost the argument.
- Praise in public, correct in private. Don't ever, ever publicly embarrass, correct, or admonish anyone. You'll create a mortal enemy for life.
- Leaders understand that you have to act, speak, dress, and behave like

a leader every hour of every day. It's not something you turn on and off, depending upon the occasion.

- In any unpleasant or difficult matter, a horrible end is better than horror without end.
- Nothing ever gets better by itself; don't ignore a problem. It will grow to be unmanageable. Better yet, seek out or anticipate problems to give you some time to solve them.
- When something goes wrong, stand up and take the blame first. No one knows what to do or say after that.
- The shortest and least significant words in the English language are "I" and "me."
- "Never mistake activity for achievement." —John Wooden
- No one works *for you*. They work *with you* or they work *for the department*.
- Never make a "gut" personnel decision. Do your homework, get some help, take your time. Hiring or promoting someone is a decision with a 25-year lifespan.
- How is what you're doing right now going to look on the front page of the newspaper? There are no secrets anymore; you will get caught and be exposed.
- No custom IT project ever came in on time, within budget, or functioned as planned. Buy something off the shelf and modify your work practices to adapt.
- Once or twice a month, sit quietly for 10 or 15 minutes and ask yourself, "Am I doing my best? What could I have done better? Who could have done it better?"
- Organizing around "people problems" never, ever works. Deal with the problem.
- Every fire department "rule" has somebody's name on it. You'll be much more successful as a manager if you take on the problem person rather than making another rule for the 1,000 non-problem people.
- The Three General Rules of Successful Public Administration: 1. If you're thinking about taking "the top job," never follow a hero; follow a jerk. 2. Don't ever say or do anything against animals. 3. Don't ever get involved in a public art project (e.g., designing a statue, decorating City Hall, choosing colors, deciding on museum displays).

- You can learn the important lessons in life on a scholarship (the easy way, learning from others) or you can pay the tuition (the hard way, by your own painful mistakes). To the extent possible, do your homework and get that scholarship.
- Onieal's Law of Organizations, Motorcycles, Computers, Airplanes, and Boats: "They're never big enough, they're never fast enough, and you can't afford to keep up."

Chief Onieal with (then) Captain Darren Rivers, now the chief of the Jersey City Fire Department, and (then) Firefighter Fran Reynolds, now a captain.

POLITICS

- Politics is simply the art of influence. You practiced politics on your parents when you were a child; your children practice it on you. Don't believe me? Watch a child walking with parents in a mall or an amusement park. You exercise influence every day of your life whether you're a firefighter or the chief of department. A firefighter or officer ignoring politics is like a drowning man ignoring water. Eschew politics at your peril.

- The last day you'll spend in the fire department is the #2 job—whatever it is. Once you get that fifth bugle, your new best friends are the mayor, the city council, the budget director, the personnel director, and the thousands of other people who know how to do your job better than you. Just like you did when you were coming up, everyone knows how to be the chief. Smile.

It's hard to believe, but elected and appointed officials have no clue what all that gold and all those bugles mean. They are looking for well-researched, documented, reasonable solutions to problems. If you think that the answer to every budget threat is "babies will die," you don't have a clue!

STEVE PEGRAM

From Brooklyn to the Rural Countryside

While I grew up on Long Island, and in many respects, still consider it home, I was not the most popular kid amongst public school administrators. As a matter of fact, I was not particularly welcome by them, as my only priority seemed to be the local fire department—Manhasset-Lakeville, Great Neck Vigilant and Alerts. When the horns blew, all I wanted to do was leave…so the school administrators helped me with that goal. Permanently. Thank you. My folks worked hard to find me a new school, and I landed in Pennington, NJ—with a fresh new start. No more worries about me chasing ambulances and fire engines. Until the local sirens started to blow and I headed right out the door. And did so over and over. I then joined and was ALLOWED to leave class. Paradise found. Somehow I managed to graduate—I believe it may have had something to do with my folks and some kind of donation—but that's urban legend, and Pennington was a great part of my life. That's how I met Steve Pegram. Fifteen years later—I was in for a high school reunion and this skinny kid was washing the fire engines. Seems that he was as gung-ho as I was and had a similar prior history with his previous Brooklyn public schools. From that day on, our friendship grew. Together we have made runs, crawled a few halls, built a municipal fire department from scratch (really!), fought with city hall dwellers and commanded fires. Today our friendship includes kids, grandkids, and we still go to fires together—he as the chief of neighboring Goshen Township and me as a DC at Loveland-Symmes. While I have included many metro and large department leaders in this book—and the information they pass on applies to us all—I also felt it was important to provide perspective from small town chiefs in rural and suburban areas. Steve has done quite a bit in his years, climbing up the ladder from firefighter to chief—on both the career and volunteer side—and because of that, what he has to pass on is certainly worth listening to.

The Smartest Guys in the Room

Since joining the fire service 25 years ago, I have learned a lot, and not always from the people I expected.

The biggest lesson both in life and in the fire service is that those around us, and especially those who came before us, really do know more than we do. Just like teenagers who one day tell their parents they are wrong, stupid, or don't know what they are talking about, later in life, usually about the time they have their own kids, realize Mom and Dad were the smartest people ever, and they probably should have listened a bit more. The people around you probably fall into the same category.

I was mentored early on from some *great* and well-known chief officers like Billy Goldfeder, Jack McElfish, John Buckman, Rich Marinucci and Don Manno (RIP Donnie!). They taught me how to be a chief. Then there were less-well-known chiefs, instructors, and friends who I met early in my career who shaped me the most as a firefighter, company officer, chief, father, husband, and friend.

Chief Steve Pegram, his wife Mollie, and their kids Jack and Riley

Many don't even know the influence they have had on my life and career. People like Gene Schooley, Phil Layton, Ralph Brandimier, Kevin Reading, Larry Omland, Chico Marcianti, Melvin Bryne, Eddie Kline, Otto Huber, Eddie Buchanan, Shane Ray, and Danny Corder are just a few mentors/friends to whom I owe most of my successes.

What these men taught me was integrity, patience, leadership, command presence, dedication, loyalty, and most of all looking out for their men and women. All of them are people I watched during my first 10 years in the fire service, and from each of them I took a small life lesson.

Chief Phil Layton from Swarthmore, Pennsylvania, my first chief, taught me command presence. Probably from his time in an active combat unit during the Vietnam War, you always knew who was in charge, and he never had to tell you. Phil was 100% chief all the time, not in a dictatorial way, but you knew it; he walked the walk and talked the talk. More importantly, as much as he was a chief, he was a teacher, an instructor, and that is what I think makes a good leader/mentor—someone who shares his or her knowledge and teaches the next generation. Some of my early teachers were Jim Yates, John Kubilewicz, Ted Cashel, and Jack Christmas; I underestimated the lessons they taught me early on, but use what I learned from them every day.

Chief Gene Schooley of Pennington, New Jersey, taught me about keeping a list. Seems simple, but it is probably one of the most important things I learned and continue to practice; I recommend you do, too. Nobody can remember everything that needs to be done, who to call back, and so on. Early on I watched Gene constantly whip out his small note pad from his shirt pocket and take notes, checking things off as he went. This was before the days of iPhones and tablet computers. Gene ran the Pennington Fire Company with a note pad and a pencil and did a great job!

The second and maybe the most important thing Chief Schooley taught me was the importance of family. Gene had a short tenure as chief, but during his time our fire company grew tenfold, our first ladder truck, our first recue truck, dozens of fires caused by an arson spree, and the fire company's 100th anniversary were all accomplished during his short tenure. I have never seen a greater outpouring of respect for an outgoing chief than was shown Gene when he retired as chief in 1992.

It was because he was a true chief; he led by example and gave 110% with no ego. But at the end of those busy years, he realized he was missing something very important—his kids. His kids Zach and Katie were growing up fast, and he was missing everything, so he refocused and went home. I am proud to say

Gene Schooley was the first chief to promote me in 1991, the chief to nominate me to be a chief officer in 1992, and remains a good friend and mentor today.

Some of the other chiefs (Chico, Ralph, Kevin, Larry, Melvin, Danny, Eddie, Ott, and Shane) were just people I learned from by watching their actions, behavior, demeanor, and listening to them on the radio. They all had one key command attribute, and that was remaining calm. If you want to be a good leader on and off the fireground, you must remain calm. There is nothing worse than a screamer on the radio.

When a garbage collector turns the corner in the morning, he doesn't start to scream and shout because there are trashcans all up and down the street, does he? No. That's because he is a garbage collector, and that is what he expects to find. We are the fire department, and when we're dispatched to a fire, we should expect to see fire when we arrive. A screaming chief or company officer can set the stage for a bad operation right from the start.

Likewise, if you are trying to motivate your employees around the firehouse by yelling, it won't work, and you will lose respect quickly. Sometimes yelling at someone is appropriate, especially on the scene of an emergency when something very unsafe is about to happen and the message needs to be heard and received loud and clear. Other than that, calm, cool, and collected is a much better way to work.

Make sure you realize everyone is watching you, and you should be watching them. From the minute you walk into the firehouse for the first time, everyone is sizing you up. Doesn't matter if you're the newest volunteer or the newly hired fire chief, they are sizing you up. How you act, how you talk, who you talk to, all your actions and behavior on and off duty is being watched and talked about. You only get one chance to make a good impression.

Likewise you should be sizing things up as well. Eyes and ears open, mouth shut, even new chiefs. Please realize you don't know everything, and some of them, the other firefighters, officers, and chiefs do. Now notice I said some; there are in every organization people who have risen to positions of authority that don't belong there. It is important to realize who they are, but the only way to do that is by watching and listening. You may find out the best fireground leader is a firefighter with 30 years on the job, or you may learn more from the former chief or retired firefighter who sits in the kitchen and drinks coffee most of the time. Don't be afraid to pick their brain.

Finally, learn the history of your community and your department. There is a great emphasis on tradition in the fire service; learn it, maintain it, but also know how to change it or reshape it as well. Change can be good if managed correctly; change in order just to make change can be very detrimental to morale, operations, and your respect.

Firefighters and officers who know the history of their community and department are better-educated firefighters. You will impress those who have gone before you, and it will prepare you today for the eventuality that we will all some day experience, which is being the senior person. Some day you will be the older person in the kitchen, maybe the chief or training officer.

Watch, listen, and learn, then repeat!

Joseph W. Pfeifer

Leading during Extraordinary Times

Assistant Chief Joe Pfeifer is FDNY's Chief of Counterterrorism and Emergency. During his career he has commanded some of the largest emergencies in fire department history. He was the first chief at the World Trade Center on September 11, 2001, and played a major command role during hurricane Sandy in 2012. Since the World Trade Center attacks and the loss of 343 members of FDNY, including his brother Lt. Kevin Pfeifer from Engine 33, Joe assessed the FDNY's 9/11/01 response, identified new budget and policy priorities, helped overhaul management practices, created partnerships to supplement their existing competencies with new expertise, shaped new technologies for emergency response, and developed the FDNY's first Strategic Plan and Terrorism Preparedness Strategy. Chief Pfeifer founded and directs FDNY's Center for Terrorism and Disaster Preparedness. His heartfelt work and leadership have impacted fire departments nationwide—so we never forget, and we learn so much more.

Chief Pfeifer is captured making critical decisions while battling the Breezy Point Conflagration during Hurricane Sandy. (Photo by Todd Maser.)

Doing Ordinary Things . . . So Others Might Live

Over and over again, we witness firefighters running into many dangerous situations to save lives, extinguish fires, and take care of the injured. For these acts of bravery they are often called heroes. I define a hero as *one who does ordinary things, but at extraordinary times.*

In the beginning of the 21st century, the fire service has had more than its share of extraordinary events. On September 11, 2001, we watched firefighters run into the burning building that were deliberately struck by commercial airlines. With thousands trapped in the World Trade Center, firefighters with heavy equipment climbed the narrow stairs to rescue those who could not escape on their own. As they ascended the 110-story buildings, they assisted people and encouraged others to continue their self-evacuation. On January 15, 2009, U.S. Airways Flight 1549 was forced to ditch in the Hudson River; fireboats and boats from multiple agencies responded to pluck all 155 passengers and crew from the icy waters. During Superstorm Sandy in late October of 2012, firefighters battled a conflagration at Breezy Point in New York City by using floodwater from the hurricane to extinguish the flames that spread across 148 homes. And on April 15, 2013, firefighter and emergency medical personal ran to rescue those who were critically injured by terrorists' bombs at the Boston Marathon.

Throughout the world, firefighters encounter many kinds of disasters. We battle fires; perform emergency medical care; operate at transportation accidents; mitigate hazardous materials and industrial incidents; perform search and rescue at building collapses, floods, hurricanes, blizzards, earthquakes and tsunamis; and take lifesaving actions at terrorist incidents. In each of these extreme events, firefighters do ordinary things, but at extraordinary times, under extreme physical and mental stress, when people are in their greatest moment of need.

The critical question is how can the fire service adapt to this wide range of crises to allow it to better lead in times of extreme events? To meet such challenges, firefighters, company officers, and chiefs need to develop high performance skills that enable them to adapt to novel and complex dangers.

Firefighters use the skills listed below to hastily connect to each other as well as other organizations for collaboration, coordination, and incident management. For example, after gathering situational awareness, a ladder company connects with and engine company to coordinate ventilation; the operations sector chief collaborates with rescue companies and medical personal about searches and patient care; while the incident commander coordinates with a police chief about site security and street closures. For this to occur, first responders must hone their skills so that they can adapt to different kinds of disasters.

High Performance Adaptive Skills for Extreme Events

Situational Awareness Skills
> The ability to search and make sense of information about fires and emergencies.

Networking Skills
> The ability to connect to a diverse group of people and sources of information at, and away, from the scene of an incident to develop a comprehensive operational picture.

Negotiating Skills
> The ability to collaborate with others whenever objectives or solutions cannot be achieved alone.

Decision-Making Skills
> The ability to quickly combine intuition with analysis to choose a course of action under stress and uncertainty.

Envisioning Skills
> The ability to anticipate and conceive ways to customize and improvise procedures to get things done.

Coordinating Skills
> The ability to recognize interdependencies and work with others to adapt to new threat environments.

Communicating Skills
> The ability to receive and convey timely voice, video, and data information throughout a crisis.

Managing Skills
> The ability to systematically organize and carry through with decisions, yet remain flexible for the unexpected.[1]

This mural hangs in the Fire Department Operations Center, which depicts the many different types of response by the FDNY. (Photo by FDNY.)

At extreme disasters, firefighters and other first responders are expected to act decisively under conditions of vast uncertainty and perform complex tasks to protect life and property. Yet, there is no greater stress than making critical decisions in dangerous situations. Making decisions at fires and emergencies requires the ability to trust one's instinct, but also to stop and think about what to do. Knowing when to rely on one's gut feelings and when to switch to analytical thinking is vital for firefighters who run into burning buildings and for chiefs who are in command positions.

On September 11, 2001, small decisions became the difference between life and death, which made it important for firefighters to blend rapid and deliberate thinking.[2] At 9:59 a.m., inside the North Tower of the World Trade Center, there was a loud roar and the building shook. In a fraction of a second, we knew something was very wrong and moved quickly from the lobby to an adjacent passageway. This gut feeling or intuitive thinking was not generated by the knowledge of the collapsing South Tower, but instead by matching the sound to similar experiences of a local collapse. Interpreting this roar as a dangerous condition at the lobby command post, we moved swiftly from the falling debris. Shortly after the rumbling stopped, we were covered in chocking dust and complete darkness. This rapid thinking and decision making saved our lives and bought us time to think about what to do next.

In the darkness of the North Tower, some chiefs correctly continued with rapid thinking and wanted to find a way out of the building. However, having knowledge of the World Trade Center complex, I knew how to get out, which allowed me to switch from rapid thinking to deliberately focusing on selecting the next critical command. In less than a minute, it became clear that if we could not command in the North Tower , we had to withdraw our firefighters from the building. I depressed the transmission button on my portable radio and gave the following orders, "Command to all units in Tower 1, evacuate the building." While this might seem like an obvious decision for those watching videos of that day, it was not so obvious to those commanding in the North Tower, or the firefighters under our command engaged on uppers floors who did not have the information that the South Tower had collapsed. Switching from rapid to deliberate thinking is even more demanding under high-stress, dangerous conditions. Commanders had to overcome their first instinct to continue rescue operations and instead make the critical decision to abandon a burning high-rise office building with hundreds of people still trapped. Moving from rapid to deliberate thinking allowed fire chiefs to concentrate on commanding, which lead to the flash of insight to evacuate firefighters from the North Tower.[3]

We see this same type of switching between these two modes of decision making by firefighters on May, 1, 2010, during the attempted bombing by Fisal Shahzad in Times Square. Firefighters and their company officers had this gut feeling that *something didn't look right* and deliberately stopped and asked the police to run the plates of the abandon SUV at the corner of 45th Street and 7th Avenue. When the plates came back unregistered, both firefighters and police officers concluded that this could be a vehicle-borne improvised explosive device and decided to evacuate the area. It is critical for all first responders to use both rapid and deliberate thinking at fires and emergencies.

With so much danger that surrounds firefighting, why do firefighters run into hazardous situations? The usual answer given is that it is our job. However, when this question was asked to me by the loved ones of the 343 firefighters who died on 9/11, it became the most difficult question I ever had to answer. Families wanted to know why firefighters ran into the burning World Trade Center. They struggled with trying to understand how firefighters could consciously decide to put their lives at risk. The answer was right there in front of all of us, but was hidden for a while by overwhelming grief. *We ran into danger on 9/11, so that others might live.* This simple statement defines who we were that fatefully day, who we are today, and why firefighters run into peril. Developing high-performance adaptive skills will improve decision making and lessen risk, so that we can continue to do ordinary things at extraordinary times.

NOTES

1. Modified from Harvard Kennedy School's Programs on Crisis Leadership, Discussion Paper Series: Pfeifer. J. 2013. *Crisis Leadership: The Art of Adapting to Extreme Events*. Cambridge, MA: Harvard Kennedy School. http://www.hks.harvard.edu/var/ezp_site/storage/fckeditor/file/pdfs/centers-programs/programs/crisis-leadership/Pfeifer%20Crisis%20Leadership--March%20 20%202013.pdf.
2. For further discuss on fast and slow thinking, which this section on decision making is based on refer to: Kahneman, D. 2011. *Thinking Fast and Slow*. New York: Farrar, Straus and Giroux.
3. For a more detailed analysis of critical decision making in dangerous situations refer to: Pfeifer, J. and Merlo, J. 2011. *The Decisive Moment: The Science of Decision Making under Stress*; and Sweeney, P., M. Matthew, and P. Lester, Eds. *Leadership in Dangerous Situation: A Handbook for the Armed Forces, Emergency Services and First Responders*. Annapolis: Naval Press. 230–248.

Shane Ray

Don't Let That Slow Talker Fool You

Shane Ray's southern accent can fool you into thinking he's laid back, but nothing could be further from the truth when it comes to preventing fires and minimizing injury and death. In recent years, especially since he became South Carolina's state fire marshal, he is leading the way in figuring out how to get the fire out as soon as possible. Deck guns? Yep. Hand lines? Yep. Fire sprinklers? **Hell yeah.** And he is definitely getting some attention. Shane began his career in 1984 serving every rank from firefighter to chief. He has stood out as a bold leader in regard to the need to prevent fires and, when we can't, getting water on them as soon as possible, and he really doesn't care how the water gets on it, just get the fire out ASAP. Interesting concept from an equally interesting guy who has a lot to pass on.

A thoughtful Shane Ray is always looking for ways to "do it different"—whether in the classroom, in the field, or in the office.

Do It Different . . . Don't Risk It All!

What is *it*? What is *different*? What is *risk*? What is *all*? How do *you* determine? How do others view the risk you take? What influence do you have on others? Is your perspective correct? Are your actions safe? Should you wait 20 years into your career to have an "aha" moment?

It would have been nice to have had a company officer or a chief officer give me a new perspective when I was first promoted to company officer. Jim Collins and Morten T. Hansen dedicate a chapter in their book *Great by Choice* to the "return on luck." The fire service should capture stories from the station bays and kitchen tables on our return on luck. I know I am blessed to have had a lot of return on luck, and although the near-miss reports aren't as highly reported as they should be, they are frequently told in stories.

FAST WATER

Our constant thoughts should be on preventing the fire and how best to apply education, engineering, and enforcement to our communities. However, when the fire does occur, our efforts must be on applying water as fast as possible. As fire officers and firefighters our focus has to be on minimizing risks to all, because that's the oath we took when we pinned on our first badge—to save lives and property. I wish someone would have provided the perspective of *fast water* to me 20 years before I had an "aha" about it!

The current and future built environment we operate in is much different than the environment within which our training and deployment methods were established. The built environment is larger, tighter, and contains more contents than ever before. We are fighting a different fire today, and we knew this before scientific fire research confirmed it. Why didn't we take action to change our training so we could change our strategy and tactics? Should we?

Another constantly changing environment is the political environment, which has been and probably always will want to limit taxes and the size of government. We should be aware of our environment, lead the improvement of our profession, and most importantly continually improve how we serve the citizens. We are here for service, not for self. If we lead how we serve—focused on the most effective service at the most efficient utilization of resources—we won't always have to be defending past practices that often haven't evolved with the new environment.

The fastest way to apply water to a fire is to have a firefighter with water closest to the fire when it starts. The most effective and efficient way to do this is with fire sprinklers. The fire sprinklers should be viewed by the fire service as the initial attack crew and not in a negative manner. The initial attack crew always needs backup.

In buildings not equipped with fire sprinklers we should use our technology, tools, and training to do what we learned in week one of firefighter training: limit the oxygen, reduce the heat, and prevent the fuel from off-gassing.

As firefighters, we were called to extinguish the fire, not to be entertained by it. As some of the old training books said, we came to "kill the fire" and we should do that before it kills any citizens or us. Fire sprinklers are the best way to ensure no one is killed, and our nozzle and our training are the best means to aggressively attack a fire from a safe position. The fire sprinkler is close to the fire and our nozzle allows us to be far from the fire.

THE E'S OF OUR OATH

- Education: Ourselves first
- Engineering: Use codes to create a safer environment
- Enforcement: We should consider ourselves fire prevention officers first

---------------------- **FIRE OCCURS** ----------------------

- Early warning: Smoke alarms—those capable of getting out, should; or close doors to isolate themselves from the fire.
- Early suppression: Fire sprinklers are the initial attack team.
- Emergency response: Our quick action will always be needed; the quicker we can deploy, the better we will be able to fulfill our oath.

The time is now for the fire service to "do it different." The world is looking for leaders to serve in a manner that transitions us from the old to the new. There is a need for leaders who can develop and implement a new perspective to old and continual problems. The fire service produces the best problem-solvers in the world; however, we also tend to hold onto the past more than we should. We have t-shirts, tattoos, and window stickers that say "Never Forget." It seems to me we have a hard time remembering, because we more often than not repeat the actions that cause us to have to remember more. We need to add some words and take more action. Never Forget the Fallen. Never Repeat the Condition.

We can change the conversation at the kitchen table and build upon what we know. We must build a team of leaders who can and will accept the challenge. "Do it different" offers us the answers to many of the questions that I posed in paragraph one. Let's work together to create a safer, better world.

Jay Reardon

Mutual Aid Defined

When I think of mutual aid I think of MABAS—the Mutual Aid Box Alarm System that started in 1968 in Northern Illinois to provide a swift, standardized, and effective method for assistance at extra alarm fires and mass casualty incidents. While he would be the first to tell you that MABAS is a cooperative effort, Chief Jay Reardon is the rudder that keeps it moving forward in coordination with the MABAS executive board. You think it's challenging for your neighboring fire departments to get along? Try doing it with several states. Actually, he makes it seem easy. Jay "gets" the concept of doing what's right for those who need help in an organized, fair, and non-ego-based system. When you are able to keep this many departments focused on "the ball" of doing what's best for those needing help, there is a lot for the rest of us to learn.

Jay Reardon

Promising People

What makes a promising person in the fire service? Perhaps a promising person in our profession is the individual everyone wants to listen to—when they hold court, everyone gives them their attention because they want to hear what they have to say. History has demonstrated the individual's propensity to share things of value. What they say makes sense—they demonstrate a proven credibility and rational course of logic.

The U.S. military academies have mission statements regarding their reasons to exist. All are dedicated to the mission of producing military professionals and leaders prepared for tomorrow's challenges—promising people are leaders. Military leaders possess a number of personal qualities, such as competence, confidence, compassion, humility, and integrity. All are easy to say, but it takes a lifetime of commitment for these qualities to become part of your being.

Leadership qualities as noted by our country's military academies can be transferred for fire service applications. For company or battalion leaders, here are some targets.

Competence can best be described as knowing your job, your role, and how you contribute to your organization. Your technical knowledge base should be recognized, and your colleagues consider you their go-to person for technical direction. You always have the practical ideas to solve a problem.

Confidence is best described as believing in yourself and displaying a command presence when others are pinging like a well-hit golf ball. You are not cocky or in love with yourself, but rather have a long-term record of demonstrating sound decision making in the past. Your confidence has become an indicator of success for the future in the eyes of your coworkers. You have demonstrated a willingness to take on leadership responsibilities, and you have a demonstrated ability to create or have vision. You demonstrate a confidence in your team—you've trained them.

Compassion can be easily confused as a weakness—often interpreted as the ability to cut someone some slack. Compassion is not that at all; compassion is the quality of understanding difficulties faced by everyone going through life with an added touch or two. Compassion requires some mentoring skills and a candid ability to articulate sometimes difficult words to another about their shortcomings. Telling the truth isn't always easy, but as a leader being honest and straightforward is critical when another's behavior or talents fall short of the expected standard. Someone who has a dream but lacks the talent needed to make a dream the reality needs to know what it will take to succeed. Effective mentors require compassion for the purpose of helping others develop career and life plans. Effective leaders know how to compassionately tell someone they lack critical abilities and talents. They have an above-average ability to sustain difficult and grueling mental challenges. Compassion sometimes requires telling someone they are in the wrong line of work.

Humility requires that you haven't forgotten where you came from and that you have an awareness that others allow you to lead. You would not have succeeded if it weren't for the efforts and sense of completeness your team invested in your success. Humility requires you to recognize your team as talented individuals who get the job done. Your commitment to your staff should be through your formal and informal actions recognizing members of your team—don't worry about yourself. Believe me; your demonstration of humility by focusing on others will build a loyalty from your team that's priceless.

Integrity requires you to do the right thing at the right time for the right reason regardless of whether everyone or no one is looking. People who focus and lean on the status quo are not leaders. A leader who sees problems and pursues their resolve is appreciated by their bosses, peers, and subordinates. Respect for everyone is part of integrity not just at the firehouse, but also at home with your family. It all adds up; if you invest in yourself, others will be willing to invest in you. If you invest in others, the dividends are plentiful.

Who are promising people? They are leaders in the fire service. Personal leadership qualities involve competence, confidence, compassion, humility, and integrity. Behaving with a mentor's mindset will help you keep your personal compass heading in the right direction.

Your fellow firefighters would rather see their officer behave like a boss—a boss with leadership skills. Friendship usually follows. Peer respect demonstrates character and personality, building those friendships for a lifetime. What will be your legacy? Everyone in the fire service is given a reputation—what do you want them to say about you when you retire?

Good luck!

Frank Ricci

Courage That Goes Beyond the Fireground

Frank Ricci is the drillmaster for the New Haven (CT) Fire Department and a contributing editor for *Fire Engineering*. The case of Ricci v. DeStefano was a contentious decision by the U.S. Supreme Court concerning the 14th Amendment of the Constitution and Title VII of the Civil Rights Act of 1964. Twenty city firefighters, nineteen of whom were white and one who was Hispanic, brought suit without knowing where they placed on the exam. The city invalidated the test because not enough blacks would be promoted. In June of 2009, the Supreme Court held 5-4 that New Haven's decision to ignore the test results violated the firefighters' civil rights. You may or may not agree with the basis of the suit—that's not my concern. My concern is that whether a firefighter, officer, or chief, you need to display courage for what needs to be done in the firehouse and on the fireground—and for what you believe is right. Frank has displayed this courage and shares his advice for us.

The Four Horsemen is a Paul Combs original from Frank Ricci's 2010 FDIC keynote speech on political courage, depicting the four greatest threats to our service.

The Mission, the Men, and Me

Our society and service excels when our members put the mission first. Leaders and chiefs must understand that in order achieve the mission, the troops must be trained and provided for. Some say that today's society and our members are too self-focused. I disagree; accountability starts with you and your commitment to your community.

TO THE CHIEFS

If you lead, we will follow. When dealing with the troops and politicians, remember to always advocate for your service. It is your role to set the standard and to have the courage to lead. Remember, it is your responsibility to provide for the troops and be accountable for their and your own actions and inactions.

AT THE FIRES

If your department is responding with fewer than 24 members to a two-story dwelling then you are failing as a chief and a leader. I often hear chiefs and company officers complain about staffing, yet they lack the courage to pick up the mic and call for the resources they need to coordinate a fire attack and have the necessary reserves to answer a Mayday. Note: There are no politicians on scene to say you can't call for the help you need. You are in command, so *command*.

Ensure that your bosses and chiefs know who is on scene and where they are operating. Utilize and maximize your command staff. As Chief Avillo says, "Break up the opinion brigade and delegate chiefs to areas that the incident commander cannot see." Keep the RIT boss informed and working on proactive actions that do not require the use of air (360-degree size-up, laddering, lighting, and, yes, in the north throwing down sand).

At a line-of-duty funeral you would not advocate for only two or three pallbearers to carry our fallen. So how in God's good name can you justify having a RIT team of two or three to remove a downed firefighter in a hostile environment? This is a company task of four to six at a minimum. Everyone wants to go to your fire, so if you're a small department, train with them and invite them.

IN THE FIREHOUSE

Input is free. Your members want to be involved with your vision. Whether it is a new truck, a training program, or a firehouse, ask the stakeholders for their opinion. Yes, you are the chief, and in most cases you will get to make the final call. Firefighters are not asking that you always agree with them; they are simply asking to be heard and be a part of the process.

THE POLITICAL ARENA

Every time you speak to an elected leader, start off with one positive comment about your department. Through your tenure you will have to deal with budget cuts, discipline issues, and damage control of complex situations. If you start your conversations off with a negative issue, you will be responsible for forever wounding your organization's reputational equity. Never make decisions based on emotion; focus on the circumstance and mitigating factors.

Merit must always trump politics. Ours is not a service where second best will ever be good enough. It is your job to sell your profession, and always keep in mind that recruitment of quality volunteers or hiring of quality paid firefighters will avoid embarrassment and 20-year problems. Equal opportunity for all does not equate to equal outcomes. In the real world, not everyone gets a certificate.

TRAINING

Firefighters do not rise to meet expectations; they fall to their level of training. All of our skills are perishable. Training must not be punitive; it must be fun and based on science, hard work, and proven tactics and strategies. You must get your members to buy into training. Simple skill programs that require no more than an hour a day are a great way to motivate and increase proficiency of the troops.

COMPANY OFFICERS

Set expectations, lead by example, and back up your company. Slow down and take a look. Andy Fredricks once asked, "When the trash man turns the corner, does he get excited to see trash?" You are a boss now; there is no time for emotion. Take a few extra seconds. This action will save minutes and possibly your crew's life. When faced with a non-fire dilemma, I always remember advice from the book *It's Your Ship*. Think if your actions were published on the front page, above the fold. Could you justify it, and would you be proud of your actions? If the answer is yes, you could never go wrong.

The best way to communicate with your team is through training. This will lead to less direction on the fireground. Be mindful of becoming overly tactically involved. This will cause you to lose perspective and endanger your crew. At an alarm you should be able to assign a task without having to explain how to accomplish it.

Every call is an opportunity to train, so position at all medicals the same way you position at a fire. After each fire, critique with your crew while still on scene and start the critique with your own mistakes or what you could have done better. We all make mistakes and have to make almost perfect decisions based on imperfect information. It is not about getting jammed up; it is about how you recover.

IN THE FIREHOUSE

If you are a junior boss in the house and the old salt boss above you is not as motivated, change what you can and work with your company. Volunteer to lead the drills, and communicate your vision with quiet persistence. You are not expected to change the world in one day; however, you are expected to encourage change every day.

Don't be afraid to get dirty or fail, no matter the task. Volunteer to always go first, even if you get it wrong. You are a boss; however, on occasion, wash the floor, truck, and clean with your company. If you achieved your position based on merit, then you have nothing to prove. Be comfortable in your own skin. Margaret Thatcher once said, "If you have to tell someone you're a lady, you probably are not." Be fair and have the courage to lead. Integrity is not just what you do when no one is looking; it is how you carry yourself when everyone is looking.

THE PUBLIC

Treat the public with the highest respect. There are only three impressions you can leave with our customers:

- Indifference: Nothing is lost or gained by the interaction.
- You create a new best friend: Someone who will write a letter to the editor or fight a firehouse closing.
- You leave a negative impression: They think you're an ass.
- What impression do you want to leave? Just taking a little extra time on each alarm can make a big difference.

YOUR PERSONNEL IN WORK AND OUT OF WORK

Find out their goals and help build them. Get to know your personnel and what makes them tick. Always keep your word. If invited out, show up, but leave early enough so they can talk about you behind your back.

THE NEW FIREFIGHTER

This is a dangerous job. If you do this job long enough you will get hurt. Realize the day you die or receive a life-changing injury will start like any other.

Even minor injures quite often end with surgery. With 16 years in New Haven I am having surgery for my second minor injury. The first one I messed up, fell into an open shaft at a fire, and had a firefighter fall on top of me, ripping the muscles off my shoulder. The second one, my crew and I saved a jumper and in the struggle my shoulder was torn. You just never know what can happen at work.

We must all strive to do better. It is okay to mess up; just admit it, share it, and learn from it. Regardless of rank we are all lifetime learners. Experience is the best teacher; however, only a fool learns in that school alone. Our mission matters; you are part of a family, and no family is perfect. Although our service is second to none (the public calls 9-1-1, and we just show up, no paperwork, with the only goal to help). Every family has issues and everyone at some point will face disappointment. Do not give in to those evils; they will only eat at you. We perform for the public and our brothers and sisters. Do not become indifferent or lose faith. Remember the mission; the men and me come last.

J. GORDON ROUTLEY

No Sugarcoating—Because of the Fallen Firefighters

You should recognize his name from "The Routley Report." The report was issued following the line-of-duty death of nine firefighters from Charleston, South Carolina following a flashover and structural collapse in the Sofa Super Store fire that occurred on June 18, 2007. That independent panel, led by Gordon Routley, was hired to study the CFD following the Sofa Super Store tragedy. The report clearly identified what went wrong—and why. Like Gordon, the report was blunt and to the point, identifying numerous, predictable problems that directly led to the firefighters being exposed to excessive risks on the night of the fire. There was no way to sugarcoat what happened. In this politically correct world, people are afraid to speak up. That's not the case with Gordon. In addition to the Charleston fire, he has conducted more than 30 firefighter fatality investigations. He speaks up about what he believes in so that we can learn from those who have gone before us.

The Woodhouse fire occurred late on a Saturday afternoon and involved the upper floors of a furniture and appliance store that occupied two older buildings in downtown Montreal.

The Good Old Days Weren't All That Great—Especially for Firefighters

I have begun to realize that I have been involved in the fire service for a very long time, and that causes me to think about how much some things have changed, while other things have remained pretty much the same. I sometimes find myself talking about lessons I learned at particular fires and realize that I am talking to firefighters whose parents hadn't been born way back then. Many of those fundamental lessons are so profound that I can so clearly remember the circumstances and the realization, sometimes long after the event, that the experience had been engraved into my thought processes.

I am writing this on the day that happens to be the 50th anniversary of one particular fire that had a major impact on my personal vision of the fire service, although it took several years to connect all of the dots. On that day, three firefighters lost their lives and more than 20 others were injured, many during prolonged rescue efforts. They died in a structural collapse while overhauling a burned-out four-story building, trying to salvage a bunch of washing machines and refrigerators. At the time, I was too young to be an active participant, but I was a front row spectator and knew many of the firefighters who were involved, including one of the victims. Today I can look back and put that event in perspective. During the past few days I was able to reminisce about that day with one of my friends who was directly involved in the rescue operation.

Believe it or not, our current preoccupation with safety is a relatively new development. It used to be an accepted fact that firefighters would die in the line of duty, and no one was surprised when fatalities occurred. I grew up thinking that it was sad, but inevitable, that firefighters would be killed. The old-school thinking was directed toward putting out the fire, and the firefighters were expendable—the risk of death came with the job. We developed the traditions of honoring the fallen with monuments and elaborate funerals, and we applied those rituals year after year. There were no major investigations when fatalities occurred back then—bad stuff just happened.

The fire was controlled within an hour by elevated master streams, resulting in a cascade of water onto the lower floors of the building. Moments after this photo was taken, the roof and upper floors collapsed, trapping several firefighters in the debris. Two firefighters were killed instantly. The third victim was trapped in the debris for almost six hours and died in the hospital the next day.

When I reached the age that I could actually fight fires, self-contained breathing apparatus was still a novelty reserved for special situations. (I refer to it today as "for emergency use only.") We had lots of fires and lots of personnel, but there were no rapid intervention crews, no incident management systems, no rules of engagement, and no OSHA. Chief officers developed their reputations on their ass-kicking and shouting skills, and firefighters were respected for their ability to take a serious beating and come back for more.

Many of the guys I worked with back then are dead now. Several were taken by various forms of cancer, or respiratory disease, or heart disease, or Parkinson's disease. They were great firefighters and wonderful characters, but most of them were severely beaten up and worn down by the time they retired. If they lived long enough to collect a pension, they did not get all of their contributions back. I miss them and think about how their lives would have been different if they had been born 20 years later.

Over the years we developed an enlightened system of values, and it has been decreed that firefighter safety must be managed as a top priority at every incident. We cannot remove the danger from the environment where we work, but we can do a tremendous amount to manage our exposure to the risks of sudden death, serious injury, and long-term debilitating illnesses. The responsibility to manage risk exposure applies at every level, but the higher you go in the fire service, the more personal and professional responsibility you have to assume.

The high-ranking officers who manage fire departments and assume command of major operations have a tremendous responsibility to their subordinates to ensure that their safety is always a paramount consideration. The same sense of responsibility has to extend throughout the rank structure, all the way down to company officers and even to senior firefighters who are guiding the actions of less-experienced individuals.

The ability to skilfully manage a fire scene is only one aspect of an officer's duty. We have to ensure that every firefighter who could be called upon to operate in a hazardous environment has the appropriate training and equipment, is medically and physically fit for those duties, and is operating within a system that provides every reasonable capability to manage the incident safely and effectively. These responsibilities extend to ensuring that the fire code has been enforced, prefire plans have been developed and reviewed, and every foreseeable factor has been foreseen, anticipated, and incorporated into a contingency plan. We will never achieve perfection, but we can never allow ourselves to be lulled into a false sense of security.

Now that I am one of the old guys, I can look back and see how far we have come. We have figured out how to be much more effective in dealing with much more complex situations, and we have completely changed perceptions about firefighter safety. Sometimes we become confused and think that the safety movement was forced upon us by lawyers and bureaucrats. The truth is that we matured through our own sense of duty and responsibility to each other and figured out how to introduce safety into the equation.

The morning after: all of the debris outside the building was removed by manual labor as firefighters worked for hours to extricate the victims who were trapped on the second floor.

There are still some individuals in the fire service who cling to the romantic image of firefighters risking life and limb to perform a sacred mission. Many of these proponents of living dangerously think that the emphasis on safety is compromising our fundamental values and providing excuses for the faint hearted. Few things cause my blood to boil faster than an Internet discussion forum where an anonymous individual with 18 months of experience comments on a video and refers to the participants as cowards. The time has passed where we can operate with reckless abandon and take stupid risks simply because we believe that those actions will fulfill our heroic fantasies. Now we know better and we have the ability to work with a much greater degree of safety.

Firefighters have to be prepared to work in exceptionally dangerous environments and face situations that are sometimes unpredictable. Courage is essential to meet these challenges, but raw courage is no substitute for training, physical strength, endurance, and the dozens of components of an effective system that enable us to work efficiently and safely in spite of the danger. There are times when we are called upon to risk a lot in order to save a lot, but those are rare and exceptional situations. In my many years of experience there have been hundreds of opportunities to identify and evaluate the risk factors and get the job done safely and effectively.

I sincerely hope that the next generation of firefighters will have the same opportunity to look back at us and wonder how we managed to get the job done with our crude methods and the extreme risks we had to face.

Dennis L. Rubin

Traveling the Nation—Extinguishing Fires

When you see the many places that Dennis has served, from firefighter to chief, he has definitely traveled the country and gained loads of experiences. Most recently he was chief in Washington D.C., serving from 2007 to 2010. We became friends when we went through the NFA EFO program together back in the 80s. He is a vocal, spirited man who has made his mark, and not unlike others who choose to speak up on matters important to them, he has had to knock down a few fires in his career—actual, political, and personnel. He continues to have that "fire" spirit whenever he teaches or writes and often is successful in reminding firefighters why they became firefighters and leaving them feeling good about it. Dennis is a very gung-ho and colorful man whose experiences, enthusiasm, and focus on the fire service provide him with lots to pass on to us.

The Best Discipline Is Self-Discipline

Being a firefighter (both volunteer and career) has been "living the dream" for me. The fire-rescue service has been the major and central part of my life since the ripe old age of 8 (I know, I know, I started rather late to become a junior fireman—I still have that tiny BVFD breast badge in my memorabilia collection). I have been lucky to be a member of several great fire departments, and I have served at various ranks. I have met some of the very best and brightest, liked-minded fire service individuals. All were just as proud to serve in their communities as I was. What an adventure the fire service had been so far! My membership has been worth holding on to, and I have always felt a need to be an active part of the fire service. Every minute of my experience has been without any regrets. I am perhaps the luckiest and happiest person that you will ever get to meet!

In stark contrast, I have interacted with my share of folks who were poor to marginal in the performance of their duties. Most were "misfits" and not cut out to be a part of America's bravest. If the department does not select high-quality people for entry into the fire service at the beginning of a career, the workplace problems will turn out to be poor personal behavior and judgment by the questionable member. The folks who come up short are the members (and in some cases former members) who have misbehaved either on or off duty. My advice to everyone is to always remember: *The best discipline is self-discipline*, and *behave at the firehouse and at home.*

The very day that we are fortunate enough to pin on the firefighter's badge, the public sincerely trusts us. In this day in age when no one trusts or believes in just about all levels of government, holding the public trust is quite remarkable. Most community surveys reveal that the fire-rescue department is the most trusted and respected service provided within a jurisdiction. The positive rating is typically well over 90%, without any other government agency even close on the customer service survey scale. When a citizen or visitor in your community is having the worst day of his or her life, Big Red carrying our troops into battle will be a welcome sight. Over a career, firefighters will be asked to help distraught citizens in more ways than can be imagined. As an example, our firefighters will be asked to assist a young person (say a

16-year-old boy or girl) who has received traumatic injuries in an auto crash. By the very nature of the injuries, and because this procedure is spelled out in our treatment protocols, the firefighters will be required to remove (likely by cutting them off with shears) all of the patient's clothing. The purpose of exposing this person's body will be to careful look for other injuries and to treat the once-hidden wounds or immobilize broken bones. The only way that the injured person's family and/or their community will allow you to care for this patient in such an invasive fashion is to have and to always uphold the "public's trust."

Three DCFD Recruit School classes starting their career journey in the summer of 2009

On some other day, a home in the community will catch fire. The responding firefighters will once again be a welcome sight. They will be called upon to stop the raging blaze from destroying the entire building. No one will ask the firefighters for identification or check their pockets when they exit the damaged dwelling place. The firm belief is that every one of the customer's valuables will be located in the same place it was in before the inferno started. We are only welcomed in this great moment of stressful disaster only because of the "public trust" that the fire department has earned. Our fire department predecessors, over decades or even centuries of faithful service, earned the trust of our communities. Oddly enough, this mission-critical "public trust" is very fragile and can be destroyed quickly by poor (on- or off-duty) behavior by our membership. The reasons are obvious why the community will not let questionable-to-nefarious persons pin on a firefighter's badge. Why extend the public trust to anyone who doesn't deserve it? Simply put, firefighters will not be able to perform their duties without the trust of the people who they are sworn to protect.

When a department member is in serious trouble, it is not usually because the firefighter can't deliver high-quality emergency medical care or because he or she did not understand how to advance a hoseline into a burning building. The fact is, real trouble will follow the folks who have little or no self-discipline. Always think before you act, both on and off-duty, making sure that what you are about to do is legal, moral, and acceptable. Most of the skills that are learned in the classroom are very important to enable you to perform firefighting duties properly. However, there are not many other skills that can out preform *self-discipline*. Self-discipline is a must for every member to be successful in every phase of what we do as firefighters. Most modern fire-rescue recruit schools have a significant amount of leadership (self-discipline) training added to the basic firefighter curriculum. Even at entry level, this type of leadership training pays great dividends for the new recruits and for the department. Always take a second to think before you act, measuring how your actions will look and be perceived by all (inside and outside the agency).

Not many fire officers enjoy implementing or enforcing discipline directed at another member. Trust me on this one, very few members enjoy delivering punishment to others. Always exhibit self-control and self-discipline so that you will not compel someone to enforce the rules (policies) at your expense. The organizational stress that discipline causes is significant. I was taught that when a member gets disciplined, the member, supervisor, and department failed at some point in the process. Perhaps the department failed to hire the correct member in the first place; or the supervisor failed to properly train the member to include the organizational policy, rules, and expectations; or perhaps the supervision was not appropriate for the situation; or maybe the member failed at maintaining self-discipline in the situation.

The most successful people in the fire-rescue service are the ones who are self-disciplined. In fact, people who exhibit this trait are the most successful in our society as well. Always remember that *the best discipline is self-discipline*. In departments where the members behave themselves, there is less organizational stress, happier members, and in general it is a better-managed department because the officers can focus on being officers and not wasting time on disciplinary issues that could have been avoided.

It might seem unfair that a firefighter will be punished at work for a DUI conviction off duty. Because of the public trust, we are held to a higher standard than most all other professions (or volunteer avocations). Sometimes life is not fair, and this may be one of those cases.

Be proud that you are a part of your department and the greatest profession on this planet. Do your part to make a better fire service by simply being self-disciplined and always behave yourself. If you do, a great ride is in front of you. You will be an asset to your department, and you will be much more likely be promoted to positions of higher responsibility. Until next time, be safe out there!

John Salka

It's a Fireground—Not Your Playground

After getting to know John in the late 90s, we discovered that we both started our careers as volunteer firefighters on Long Island. He was with Mineola and I was with Manhasset-Lakeville. There are many "alumni" firefighters from Long Island, a place not like many others. I once had a chance to ride with John in the Bronx, and one fire absolutely impressed me and affirmed some of my beliefs. It was a multi-family high-rise fire with the first-due company confirming a working fire. After size up one of the companies decided to freelance instead of executing John's orders. When John discovered this he immediately ordered the entire company out of the building and had them sit on the sidelines (pretty much like time out) until he was done. He then proceeded to deal with the company officer. His message: he was in command, what he said goes, and there would be no nonsense and no freelancing on his fireground. Every aspect of that incident had to be within his command, control, and accountability. Be sure to remember this story—it's a fireground, not a playground.

John Salka

Keep Reading, Studying, and Learning

I'm sure the advice, suggestions, and lessons you are reading in this book from some of the greatest fire service personalities in America today are interesting, valid, and helpful. I am going to keep my message short and sweet. If you want to be a successful firefighter, officer, or chief, you've got to stay on top of your game. If you are a firefighter, you need to know and practice and review the tactics that you are expected to perform while on duty or on a run. Every resident, every visitor, every person just driving through your community is depending on you to know exactly what you are doing. This doesn't mean read it once and you're good. It means staying current and constantly reviewing and training and practicing your trade. Constantly!

If you are a company officer, you are now not only responsible for your own safety but for the safety and survival of your crew. Oh, and for the survival of your customers, the people who call the fire department for help thousands of times a day. You need to know what your firefighters need to do their job, and you need to know what the chief needs when he gives you an assignment at an operation. You need to keep things running smoothly back at the firehouse and efficiently on the fireground. The only way to do all of this is to read, study, and learn.

If you are a chief officer of any rank, you need to stay current, too. You may stand outside at most fires you respond to these days, but what you see and what you do can have a dramatic effect on the success of the operation and the survival of your firefighters and citizens.

All of you should read a fire service magazine every month and look at what's going on in your world. Look at the newest and best ideas that are being talked about and developed. Attend conferences and seminars and listen to the folks who are there with their stories, lessons, and wisdom. It may not all apply to you, but the experience will broaden your horizons. Every day, every shift, every drill at the firehouse should be a learning experience. Don't ever go home without learning something new. Ask a question, handle a tool, or join a discussion.

Remember to keep reading, studying, and learning. You won't regret it!

José A. Santiago

Escaping the Streets . . . to Serve

In 2012 José was appointed fire commissioner of the Chicago Fire Department. He is in command of nearly 5,000 sworn members with an annual budget of over $560 million. As the only son of a working-class single mom and a father who was "never around," he dropped out of school at 17 to join the Marines (serving in Operations Frequent Wind RVN, Desert Shield, and Desert Storm SWA) and escape the street gangs that had swallowed up so many of his friends. "A bunch of my friends that I knew from school were all dead [or] dying. It was just a matter of time. . . . I had to get out of the neighborhood." He returned to Chicago in 1975 and took the CFD test and landed No. 10 on the hiring list. He was hired in September 1979, and was on the job for all of two days when he made his first rescue. Within five months, firefighters went out on strike. José was told by his union brethren that the strike would "probably last hours." Instead, it dragged on for 23 days. The rookie firefighter never once considered crossing the picket line. In fact, José and his colleagues at Engine Co. 76 spent the strike monitoring the fire radio and responding to every fire in their district. "We were not gonna let someone die in our neighborhood because of the strike," he said.

Now, as commissioner, he faces numerous challenges, particularly related to fiscal issues and progressive modernization, with the help of his decades of experience.

José A. Santiago

The Value of Leadership

The fire service has a distinct irony to it, and it needs to be addressed. We have a rank structure that is similar to the military. We stress the chain of command, span of control, standard operating procedures, and general orders.

We even call ourselves paramilitary organizations, but despite similarities to the military, the fire service misses a key element that is critical to our performance and longevity.

We take candidates and we make firefighters out of them. We tell young firefighters to prepare themselves to rise in the ranks by studying for the lieutenant's test and then captain's test, as they seek more responsibility and better pay. But, are we preparing them for the leadership role they are seeking? Are we making sure tomorrow's officers are ready to really be responsible for the health and welfare of men and women under them?

In the Marine Corps, potential leaders are identified and trained. A good officer can spot a Marine with leadership potential and ensure that he or she attends NCO school to learn the real meaning of leadership, things we don't teach in the fire academy, like human understanding, moral character, all part of the 14 leadership traits that have been identified by the Marine Corps as important to successfully lead a group to a common goal.

We must plant the seeds of leadership the moment we send candidates into the field. We cannot have them serve for years and then take a test for a higher rank, assuming they picked up leadership somewhere along the line. Training to fight fires and conduct other specialized operations *must* go hand in hand with training to lead. We need to adjust the attitude of senior members to mentor our younger members on not only tools and tactics but on attitude, responsibility, and compassion.

Our fire training will tell you how many lines to use and what rescue techniques work best. But the alarming truth is that most of our departments have little or nothing that teaches us to deal with the human issues we encounter in the field and even more critically in the firehouse.

Now, I can hear some of you saying that leadership classes are expensive and your budgets are being slashed. Compare the cost of classes with the amount some departments have had to pay out on lawsuits based on social problems, including perceived race or gender issues that in many cases can be traced back to a lack of understanding for those who are from a different culture or background.

More and more our departments are made up of a better cross-section of people, but are we teaching our people to be able to lead people they may not have grown up with? Who they may not completely understand?

If we analyze the time our members spend together, the vast majority is firehouse time, which is a social setting, not an emergency scene. We need to be better able to deal with the human dynamics of leading and following, something most departments have at the bottom of the list, if it's on it at all.

We also need to be able to spot those who show leadership qualities and allow them to emerge. But that is not to say that all of our members should not undergo leadership training, because there are no losers if all are trained. Those who do not emerge as leaders will benefit from understanding the culture of leadership that can stop kickback against those who do become leaders. In short, we can all learn if we are all exposed to leadership training that makes us proactive rather than reactive. It helps us create a culture of understanding.

While I point out how the military does this, we can look even closer to home for some clues, ideas, and programs. Private industry has long known the wealth that can be gained by having employees work to their maximum potential—a potential best realized if they have a better understanding of human dynamics.

Even municipal government has realized the need. Many cities hold leadership training for employees who show leadership potential or others who find themselves in a high-ranking position. Classes come under various names such as "executive development" or "advancement preparation," but in the end they are basically the same thing, teaching how to work in a group, how to lead that group, and how to understand that it's easier to lead people who understand and respect the methods you use.

As our older members retire, we are advancing more and more younger people into command tasks. Are they really ready? Leadership training not only needs to include lessons on human understanding, but we need to also tap the knowledge of those who have already, or who are about to, retire from this job.

A battalion chief who leaves the job after 30 years can be a wealth of information. He or she has probably worked with a good variety of people and has learned a lot along the way. But we can no longer allow raw time to do the trick. The stakes are too high. We need to institute planned leadership training as part of the regular advancement procedure. It's not a luxury in today's high-pressure society. It's just smart from a performance and economic viewpoint. Trust me. You are going to spend the money, if not on leadership training, then on legal fees, settlements, or even avoidable LODDs. Without a doubt we have some great people in the fire service, people who are genuine in their desire to be ready to unconditionally help anyone in need, at any time, at any place. We owe it to them and the fire service to make sure our members are not only physically equipped but mentally ready to lead or to follow. The choice of how you spend your money is yours, but the best decision is obvious.

Selena "Juice" Schmidt

Struck at a Scene—A Near-Death Survivor on a Mission

On April 10, 2005, Texas paramedic Selena Schmidt was slammed by a car traveling at high speeds on a rain soaked highway and sustained a broken back, head injury, and numerous severe internal injuries while she was providing medical care to a teenager in a car that slid off a slick road in Fort Worth. After numerous surgeries, months of rehab, and emotional recovery, "Juice" decided to go back to work in EMS in August 2006. So what happened her first shift back? A drunk driver hit the fire apparatus on the first call she worked on. Did it shake her up? Ya think?! But she kept on working and today continues to work as a 9-1-1 street paramedic. She also continues to educate us nationally on roadway survival by using herself as a living example of a "close call," reminding us just how close to death a paramedic or firefighter can come. Her great attitude inspires those who know her, as "Juice" passes on her experience so that none of us ever have to repeat the history that she barely survived.

A proud moment for Paramedic Trevin Bradshaw and Paramedic Selena Schmidt (son and mother) working their first shift together as partners on the streets of Arlington, Texas

A Mother's Legacy

One very early morning in 1982 when I was 11 years old, my mother's legacy was born. It was a legacy that would change our family history through several generations and affect so many lives forever. It was cold, rainy, and still dark outside. My mom grabbed my sister and me out of bed swiftly as if the house were on fire. I remember her grabbing blankets and pillows as we headed to the car. My sister and I curled up in the back seat, still in our nightgowns, half asleep as we rode in the car to somewhere nearby. Suddenly my mom stopped the car and said, "Stay here," as she walked away urgently. As I looked out of the rain-soaked window, I saw several people, a lot of pretty flashing lights, and a quick glimpse of a car that was a mangled mess. I remember not being scared as I waited and watched bright-eyed. As she returned to the car I asked, "Where did you go?" she replied, "I had to help somebody that is hurt but we can go back home now." Without even knowing it at the time, this is the moment that set the course of my life. It is the most vivid memory I have of my mother serving in emergency medical services. She was saving lives and doing it as a volunteer in the small town where we lived. She was volunteering on top of working two jobs and raising my sister and me. She is a true hero. On June 30, 1988, my mother became an emergency medical technician and was finally living her life doing something she loved, something that she was meant to do. I still had no idea what she actually did because she didn't talk about the details around us. Tragically, her mission and career were cut short only six years later when she died suddenly from a brain aneurysm. But her legacy was destined to continue.

Several years after my mom died, I made the decision to continue her mission. Being married and having two small children, I knew it was going to be tough, but I was determined. Four years later I was a paramedic and working in EMS in a large metropolitan city.

I currently work as a paramedic in private-sector EMS where we provide the 9-1-1 response for the city of Arlington, Texas. We work very closely with the city's fire department, and although we aren't housed with the fire department, we pride ourselves on the fact that we respond, train, and ultimately work together as one system. This increases the quality of patient care, safety, consistency, communication, and overall public service. I also volunteer as a tactical paramedic for a local police department. I began my career volunteering full-time at my local fire department, and I developed a profound respect for the fire service. Although those were some of the best moments of my life, my natural aptitude for medicine took me back to my roots riding the "box." My future is immeasurable, and no matter where life takes me, public service is where my heart is.

Thus far, my EMS career has not only been breathtakingly gratifying; it has taught me who I really am and about the person that I will never be. I have been empowered with the ability to cope and understand that tragedy will happen, but it is the bravery of trying to alter the outcome that makes the difference.

Throughout my career, I have become fanatically compassionate about safety. *Safety* is such a nebulous word. During my EMS education and the first few years of working on the streets as a paramedic, "scene safety" was simply a pneumonic chant that was driven deep into our brains to ensure that we didn't forget to put on our gloves and to watch out for the man with a gun . . . "BSI, scene safe!" I was never provided any formal scene safety or scene awareness training and will admit that the very function and meaning was sort of an afterthought. Sure, we learned the basics and the obvious hazards but never applied any practical scenarios or reviewed any case studies. It was just something we didn't think about, yet we work in one of the largest and busiest cities in America.

An entire new meaning and depth would be brought to the words "scene safety" on April 10, 2005. Tragedy struck while our entire crew was working at an accident scene in the center median of a busy interstate. As we were all huddled at a crashed car, another vehicle lost control in the rain, hitting us as it slammed into our scene. Unaware and complacent to the fact that human error will always pose the biggest risk to emergency responders, I never imagined that such a thing could happen. Something like this hadn't happened to anyone in our system during my tenure; I had never read about it and certainly wasn't warned about it. My mentality was that *surely* everyone will clearly see the big vehicle with bright flashing lights. I had to fight through what was now my own tragedy when I almost lost my life that day. I triumphed through months of pain, doubt, fear, and questioning my mission, my destiny, and everything that I believed in. Post-traumatic stress (PTS) is very real and fervently sneaky as it leaches itself into your dreadful waking moments and into your few sleeping ones. Shortly after my horrific incident, Chief Billy Goldfeder contacted me in an effort to warn and teach other emergency responders through sharing my experience and telling my story. I had no idea how this one phone call could affect my future as a paramedic. I attribute the future safety of myself, my crews, and my patients to Billy; where everyone else had failed me, he extraordinarily bloomed forth the message in an epic display, transforming me into sort of a safety soldier. Ignorance may not be curable, but it's at least treatable with strong, bold education. In addition to teaching emergency responders the significance of scene safety, educating the public about the dangers we face while helping others is the next important element in winning this battle. Part of my future mission is accepting this challenge.

Despite the fact that fear still haunts me in the deepest of moments, I use the strength of my family and friends to embrace it. What my eyes have to see and what my hands have to touch will forever create tiny scars in my soul. At the end of the day what matters most is that I empty my cup and find peace that I get to return home. The positive outcomes along the way make it all worth it and happiness achievable.

As my mother's legacy continues, an entire family of public servants is born. Proudly wearing the patches of EMTs, paramedics, and law enforcement are my two children, my sister, my daughter-in-law, my husband, and my very best friend, symbolizing an honor to love our work, even when it doesn't love us back. We offer each other a unique support system and live by the importance of sisterhood and brotherhood, self-expression, altruism, empathy, justice, and comedy.

Never say "I can't" or you simply won't. Stay alert and positive in anything you do. The power of concentrating your whole body, your entire nervous system, your adrenaline, and all of your will on a single goal is an almost unbeatable concentration of force.

WILLIAM SHOULDIS

Positive about the Seriousness of This Job

I met Bill while attending the National Fire Academy in the early 90s, and it didn't take long for me to understand that I just wanted to learn from this man. Bill is retired as deputy chief from Philadelphia but, like others in this book, he has never stopped learning. He is an adjunct instructor at the National Fire Academy and the Emergency Management Institute. As another chief who understands the critical importance of higher education, he has a bachelor's degree in fire science administration and a master's degree in public safety. If you've been to FDIC or Firehouse, you've seen Bill. If you haven't, "bucket list" him so that you have a chance to listen. A friend of mine says she gets a real kick out of it when I describe a person (man or woman) as "a good guy." Yeah: *good guy.* It's not a term I toss around easily (although this book could be titled something like that, describing my contributors). Bill defines what I mean. His Philly attitude and genuinely positive veteran firefighter spirit is infectious, and one that is passed on when you meet him and when you read what he has to say.

Incident commanders must be prepared to switch the mode of operation based on status reports and visual observations. An offensive mode is not an obligation, but an option. (Photo by Greg Masi.)

#1 Meridian Plaza—A Story of Tragedy and Change: A Night of "High-Rise Horror"

The fire service has always been a reliable responder despite frequent downsizing. No plan is ever perfect; yet when citizens encounter an emergency situation, they merely pick up the phone and dial 9-1-1. The traditions, principles, and partnerships of first responders provide the directional arrow for fast action. The modern mission of the fire service is to deliver a trustworthy, coordinated, sustainable, and safe all-hazard service. Today, Presidential Policy Directive #8, titled "National Preparedness System," is one of the guiding beacons for consistent action. The National Incident Management System (NIMS) is another roadmap based on best practices. Both documents provide a framework for field commanders. Both encourage an integrated, not independent, approach to resource management. Fire officers are the lifeline between success and failure. They have the direct responsibility, in a dynamic environment, to predict problems and determine safe solutions. Identifying likely threats, accurately assessing risk, understanding organizational capabilities, controlling cost, and having the courage to make difficult decisions are all navigational tools for peak performance during an emergency. Preparedness is the key service and safety in any size community.

One path in the professional development journey is to examine extreme events. Analyze large-scale incidents that have forced the fire service to make operational changes.

It was a cold winter's night in Philadelphia. Most homes had heaters working overtime. In the heart of the downtown high-rise district, the flames from oiled-soaked rags were generating heat. Eventually eight floors of a high-rise building would burn out. The heat would destroy a noncombustible fire-resistive building. The details and documentation from the deadly #1 Meridian Plaza incident led to changes in several of the census standards produced by the National Fire Protection Association (NFPA).

Legal and legislative implications from this 12-alarm fire caused nightmares in the private and public sector. For me, as a battalion chief, it resulted in better appreciation of organizational policies, the requirement to streamline operational procedures, and the duty to develop to providing practical tactical/task level training. For me, the #1 Meridian Plaza experience quantified the need for first responders to have an in-depth understanding of the new technology being installed in high-rise buildings. Maintaining the status quo could not be the standard.

Upon arrival of the first units at the #1Meridian Plaza incident, fire was venting from an upper floor. At the onset, the blaze was controllable and seemed contained to a single floor. Quickly, conditions deteriorated. Initially, it was the electrical systems. The primary and secondary wires shorted. The large structure was thrown into total darkness. Handheld flashlights provided the only illuminations. Elevators stalled and restricted movement of personnel and equipment above the lobby level. Time and travel distance become major obstacles. Fatigue impacted firefighters who had no choice except to climb the fire towers stairs to the staging area on the 20th floor. For command personnel, there was an increased frustration after finding improperly set pressure reducing values on the standpipe outlets without any means to make an on-scene adjustment. Developing an effective water stream was impossible when minutes mattered most.

Meridian Plaza showing the path of destruction

Systematically the fire spread in the combustible office spaces. Soon the core temperature in the high-rise building began to rise without an operable HVAC system. Poke-hole paths in the pipe chases permitted the toxic smoke to circulate in all directions. Smoke began to bank down and the heat intensified. Without a practical way to extinguish the fire or protect the uninvolved areas, the mode of attack switched from an interior offensive position to exterior defensive posture. Manual suppression measures were stopped, and the fire spread until it reached the partially sprinklered 30th floor. There, nine sprinkler heads activated and halted the advancement of the flames.

Across this nation, the #1 Meridian Plaza experience changed attitudes—the themes of preparedness and partnership were not textbook theory. Fire officers had much to learn about the preincident planning process. Every first responder had to contribute. Data from after-action reports (AAR) needed to be collected and shared. The fuel load, construction components, and configuration of fire protection features had drastically changed without input from the fire service. The lessons from #1 Meridian Plaza were numerous. They magnified the need for private and public resources to work together.

The vision of a fire-safety community must blend high-tech upgrades and realistic training. Knowledge of active and passive fire protection features must be incorporated into strategic fireground decisions.

The three firefighters from Engine Co. #11 who lost their lives on the 28th floor of the #1 Meridian Plaza are a reminder that firefighting is a dangerous activity. In the aftermath of criticism and national feedback, the Philadelphia Fire Department had to painfully evaluate itself. A careful audit was undertaken for internal operational changes and expansion of the scope of training to include rules on air management, human limitations, knowledge of pressuring restricting devices, and nozzle selection. External changes were made with significant stakeholders. An essential element of response readiness hinges on greater involvement with building managers, building designers, qualified contracting companies, and licensing officials.

Emergency response entities that are without a prevention programs will be prone to hardships. Start an improvement process by sharing critical information, having strict compliance to codes, and creating a firm schedule for inspecting fire protection equipment.

Increase the odds of reducing the risk to first responders by establishing a working relationship with all assisting agencies. Don't wait. Look ahead. Move forward. Now is the right time to take the leadership role to pinpoint planning deficiencies so that the next generation of first responders can properly protect occupants and save our own.

William Shouldis

Ron Siarnicki

Taking Care of Firefighters: It's His Family Tradition

I met Ron in the 80s when he was working in Prince George's County, Maryland. We have remained friends ever since up to and including when Ron himself was appointed as the fire chief. After his retirement, Ron was hired as the executive director of the National Fallen Firefighters Foundation in 2001. Under Ron's direction the foundation has been re-energized like never before. Taking care of the families and friends of fallen firefighters was heavily enhanced, but in many respects, more importantly, the focus of preventing those deaths found Ron leading the foundation—and the fire service—in a totally new direction. Today the NFFF, working with its many partners such as the IAFF, IAFC, NVFC, NFPA, and many more, continues to keep death and injury prevention the basis of how we operate on the fireground and emergency scenes. Ron has received numerous awards, including the highly prestigious Mason Lankford Fire Service Leadership Award. However, he may be most proud of the fact that he comes from a long line of firefighters, as his father and grandfather both served as volunteer firefighters in western Pennsylvania. With energy like few others and a focus on what needs to be done, Ron passes on to us what he feels we need to know.

Ron Siarnicki

No One Goes Home . . . Without Being Checked Out if They Become Ill at an Incident Scene

I'm not a doctor, and I'm not a researcher. I'm just a fire chief who has, for the last dozen years, looked closely at firefighter fatality reports as part of my work for the National Fallen Firefighters Foundation. I am convinced there's a way to further reduce the number of firefighter deaths each year that doesn't require special funding or an elaborate, multiyear, double blind study.

What has jumped out at me from the pages of the reports is something very basic. We can save lives with a simple policy that says any firefighter feeling ill during or immediately after a response *must* be seen by medical personnel. No one is allowed to go home or to their bunk at the firehouse without following this step. Here's why.

Every year the death reports describe firefighters who reported they didn't feel well during or after a response. They might mention to an officer or fellow firefighter that something isn't quite right. But that's as far as it goes. They return to the station or even go home to rest, certain they'll feel better later. Unfortunately, what can happen next is they collapse and are later found dead by family or friends. This occurs as many as 10 times each year.

Let's not be misguided into thinking that this is an issue only for the veterans, those who've been in the fire service for decades. Heart disease and related health issues span every age bracket. Each year, I see younger and younger men and women succumb to health-related issues that could have been prevented if only they had talked to someone and received a medical evaluation on the scene.

It's this simple. Follow these two steps to help reverse this trend.

1. If you are feeling ill during or after a response, tell someone immediately. Don't ignore it. Don't just sleep it off.
2. If a fellow firefighter or one you supervise tells you they aren't feeling well during or after a response, require that they get medical attention. Don't take no for an answer.

I'm convinced in some of the cases reported each year that this intervention would have made a difference in the outcome for the firefighter and ultimately their family. Often firefighters are having the onset of a heart attack and need medical treatment; otherwise they are going to die.

Our bodies sometimes provide us with warning signals that we must not ignore. All too often, we make excuses. Maybe it's because of ego. Or fear of being pulled from the crew. Or we just think we're invincible. But the simple fact is an untreated coronary issue will lead to death. I believe that fewer deaths will occur each year if we admit we need help and then follow through with intervention. By taking this stand you could prevent a line-of-duty death. Making this a nonnegotiable department policy could further reduce the total number of firefighters who die each year.

In my 40 years as a firefighter, I've come to realize that all members of the fire service community have an obligation to be accountable to ourselves and to others. We also have a duty to act when we encounter unsafe actions or situations that may result in injury or death to ourselves and others. That is why we must empower ourselves and others to act when the situation warrants.

This is what I have learned. I hope you will share it.

James P. Smith

Professionalism Worth Replicating

Jim retired from the Philadelphia FD in 2007, after 41 years of service. He may have retired from his department, but certainly not the fire service. Jim has been a good friend to me and in later years, has gotten to know my kids and grandkids when we all head "down the shore" each summer. I met Jim when I was going through the Executive Fire Officer program at the National Fire Academy and was really impressed with his love for tactics and operations, but also his devotion to his own department. How often, these days, do we hear firefighters not proud of their own departments? Sure, there are valid reasons from time to time when your department isn't the greatest place on earth; that's just reality. A quintessential Philadelphian, his positive enthusiasm is as infectious as his seriousness about fireground command and operations. I could listen to Jim for hours, and have as a student. The highest civil service rank in Philly is deputy fire chief, and that's what he retired as after more than 20 years. Jim also served as director of training and as a field commander. A popular instructor at the NFA, he is the author of the textbook *Strategic and Tactical Considerations on the Fireground*. The book is essential for any firefighter's library.

What I Learned in Life and What the Fire Service Taught

My parents taught me "love and family above everything else." Mom was phenomenal. I was one of twelve children, and we each knew that we were her favorite. Dad died young; there were still six kids at home, and she taught us we could be anything we set our minds to and were willing to work to achieve.

Firefighters are also a family, one with which you often spend more time than your biological family, so learn to live with them and treat them as you want to be treated.

I remember being told by the good nuns in grade school that you have to find something in life that you enjoy doing and you will never work a day in your life. At the time I don't think that I believed that was possible. Then I found firefighting. Firefighting is a profession that I found by accident, but what a great find it has been for me. I have had some terrible days during my career involving death and suffering, but there was never a day that I didn't want to go to work. Fortunately I found that the good nuns were right.

I found that the worst job in the world is telling a wife and children that their husband and father gave his life trying to make a rescue. One young daughter challenged me with, "Why my father?" Words can never express your true feelings nor can those words be accepted by those receiving them as solace for their loss.

The most critical person in a firefighter's career is his or her first officer. That person will have the most impact on how that firefighter's career will turn out.

Though I think of myself as an excellent fireground officer, I have never had the perfect fire. After each incident, I critique myself and constantly strive for that perfect fire, and I feel that's what I need to do to keep getting better.

I have found that I learn more from my mistakes than from my successes. We should never lose sight that the goal is to get everyone home safely while protecting life and property.

The really good athletes say that the game seems to be going in slow motion. As incident commanders we need to get to that place to be really effective. We need to remember that the emergency is happening to someone else, and we are the solution to the problem.

I have always hated shouting on the fireground and thought it to be unprofessional. I would tell my officers, "Don't shout until I shout." Shouting to me always indicated an unsolvable problem, and as we know there is always a solution; we just need to find it.

Running was another thing that I didn't want to see of my firefighters at an incident scene. It always seemed that the brain tried to react as fast as the feet were moving, and that typically led to bad decision making. My one exception was when working refinery fires; if the fire brigade members started to run, I would tell my guys to follow them because they knew a great deal more about the dangers at the refineries.

Remember, heroes are ordinary people who rise to the occasion exactly when the need occurs without any thought for their own safety, but only for the safety of those needing them at that time.

Firefighting and Cancer

There have been many studies showing the relationship between firefighting and cancer. Sadly, we have all seen this too many times in the many personal experiences of cancer in our firefighting family. Even with all of the instances that we are aware of, it is always a shock when personally diagnosed with cancer.

I realized this when I visited my new primary care doctor for a minor elbow problem and then he gave me a full physical. In a consultation of some blood tests, he wanted to discuss the results of one of the tests that showed a high PSA (prostate specific antigen) count. He explained that the blood test showed an irregularity with my prostate. The doctor told me that it was probably nothing to worry about, but he wanted to do more tests. I visited a urologist who took biopsies of the prostate to rule out cancer. The only problem was that half of the biopsies tested positive. I had prostate cancer. More tests were then needed to make sure it had not spread past the prostate. Thankfully it had not spread. I had some options on how to proceed. I decided on a radical prostatectomy, which meant surgical removal of the prostate.

My urologist told me how lucky I was. I was having a hard time comprehending what he was saying since I found it pretty hard to associate cancer with good luck. I asked him; "Why was I so lucky?" He explained that his experience with prostate cancer patients was usually quite different than mine. I was 48 years old at the time, and if my primary care doctor had not given me a blood test to check my PSA, what would typically happen was that I would notice nothing unusual until I was about 55 or 56 years old. At that time, I would experience some difficulty in urinating and would likely seek medical help. The seven to eight years in between would have allowed the prostate cancer to spread to other parts of my body, and the only treatment would be to keep me as pain free as possible.

A few years later while teaching a firefighting class in Denver, Colorado, I found blood in my urine. My urologist found that I had a bladder tumor. I was treated at the hospital as an outpatient and had the tumor removed. I had a test called an Intravenous Pyelogram (IVP), which checked my kidneys. The results of the test were inconclusive, and the urologist wrote a script for me to get a CT scan of the abdomen. He was not concerned and told me it was just a precautionary measure.

I had the CT scan, and the results were not good. It showed a growth in my abdomen that was crushing my right kidney. I made an appointment and saw an oncologic surgeon. I asked about the size of the growth, and he noted that it was the size of a child's small football. He explained that it appeared to be a liposarcoma (a fatty tumor) that was wrapped around my right kidney. He also stated that he would not cut the sarcoma inside of my abdomen to avoid the possibility of cancer cells remaining inside of me. I decided to donate my own blood, which delayed the operation for about a month. The operation that was supposed to take three hours lasted six hours. The surgeon found that the tumor had grown and had wrapped around the kidney. After "unwrapping" the growth it was found that it was attached to the renal vein, which supplies blood to the kidney. The right kidney had to be removed along with the now 12-pound tumor, which had grown to the size of a regulation football.

The moral of the story is that cancer and other diseases don't just happen to other people. They can happen to you. Realize that with the exception of the blood in the urine, I had no symptoms. Many firefighters I know avoid annual physicals unless they are department mandated. I was no different in that respect. I rarely went to the doctor. I was lucky that the various cancers were found in time to prevent their spread to other parts of my body. According to the oncologic surgeon, there is no known relationship between the three types of cancer that I had. I strongly feel that my cancers are due to my many years of firefighting.

Since my encounters with cancer, I am now a poster boy for doctor visits. I have had additional bouts with bladder tumors, but fortunately no complications with the other cancers.

Learn from my experiences that you need to protect yourself at all times. Lead by example. Wear all of your protective gear. Don't let a 10-second exposure lead to a lifetime of misery. Take care of yourself so that you will be around to see your children and grandchildren grow up. Physical fitness and proper diet need to be part of our regimen. Annual physicals are another step that firefighters must take to ensure their continued health.

It is up to us to keep watch over ourselves. Like Nike says, "Just do it."

Effective Fireground Operations

How do we ensure that our fireground operation is safe and effective? The proven method is through preparation prior to the incident and a thorough analysis afterwards. During my career in the Philadelphia Fire Department I responded to thousands of fires. After each fire I critiqued myself to see what I needed to do to have that perfect fire. I reviewed my every action from the time of dispatch until I left the incident scene. I checked to see that I followed a logical approach in my thought process. I was always brutally honest with myself and tried to analyze my each and every order at an incident. If I struck a second alarm, I looked at each task that I assigned to a company. If I had a working fire and I didn't strike the second alarm, I would go over whether that was the right decision. Did I work the firefighters too hard? Should I have gotten relief for them?

When dispatched as a deputy chief I would write down the units that were responding. On arrival I would don my personal protective gear and do a 360-degree walk-around of the incident scene. I sized up the situation as if I were the first-arriving unit and determined what my initial strategies and tactics would be. I compared my observations with what was occurring on the fireground. I would then have my aide announce my arrival on the fireground.

I would meet with the battalion chief who was the incident commander. I expected a comprehensive report of his or her observations and orders given . . . the problems that still needed to be dealt with and the incident command positions already assigned. If it was a growing emergency, I would assume command.

What else needs to be done for effective fireground operations? A command post needs to be established with a view of the front of the fire building and the most endangered side. Incident objectives need to be developed, which at structural fires are based on our incident priorities. Those incident priorities are life safety, incident stabilization, and property conservation, while ensuring that firefighter safety is a prime consideration.

Based on the incident objectives, you need to develop and implement an incident action plan. That is where our strategies come into play. The four critical strategies for the initial units at an incident are: rescue, exposures, confinement, and ventilation. The incident commander needs to communicate the plan to all responders. This is done through specific company assignments. In other words the IC needs to be in command of the incident scene and control what is occurring. The implementation of the strategies has to be coordinated. Ventilation without fire attack will cause the fire to grow and threaten the life safety of firefighters and occupants. Likewise, fire attack without ventilation will fail. Problems with fire spread can occur if the initial attack is in the original fire building while the fire is spreading to adjacent exposures. This is where control by the IC is needed.

The implementation of tactics should consider where to position the first hoseline and then the second and third hoselines. It should determine what ladders should be used and where should they be placed. It is often asked if orders need to be given for each of these tactical operations. That is solely dependent upon the experience level of the responders. In many fire departments, assignments to a veteran fire officer can be broad in nature, whereas a similar assignment to a newer firefighter may need to contain specifics on what the IC requires. At every incident the IC needs to observe the firefighter's actions and receive verbal feedback to ensure that the assignment is being performed as ordered.

The IC needs to continue sizing up the incident while receiving progress reports from the interior companies, the roof, the rear, and the exposures. These reports are essential to allow you to keep abreast of exactly what is happening. Keep in mind how long you are operating, and constantly assess if any changes need to be made. Pay close attention to the reports the safety officer is giving you. Depending upon conditions decide whether you should change from an offensive attack to a defensive attack. Of course, everyone wants to put out every fire in an offensive attack. No one wants to pull out of a building and go defensive. But—and it is a big but—you as the IC, through your ongoing size-up, need to make that decision considering the safety of your firefighters and occupants.

In analyzing and investigating fire operations and doing investigative reports for fire departments, I have found some common operational problems. In many instances no one was in command of the fire. Another typical problem is due to weak command that can manifest itself in poor decision making, or no decision making on the part of the IC. The National Institute for Occupational Safety and Health (NIOSH) has many times noted these failures in their reports relating to firefighter deaths.

So my recommendation for effective fireground operations is to prepare by using a system. Read and study constantly to gain knowledge of incidents that have occurred in other fire departments. Work on your command presence at every fire. Don't assume that something is being done; it needs to be confirmed in either a face-to-face report or via radio. Observe from an established command post the actions of the firefighters and any changes in the incident.

Remember, professionalism is not something that is determined by a paycheck; it is found in the actions of well-trained and proficient firefighters.

It is all about family as we celebrate my last night in the PFD at Division 2's headquarters. Seated from the left is my son Jim a deputy chief in the Ocean City NJ Fire Department, my niece Katie O'Malley, myself, my wife and best friend Pat, and my daughter Colleen Guidos. Standing from the left is my nephew Shaun O'Malley, a firefighter in the PFD, Chris O'Malley, my daughter-in-law Kate Smith, my brother-in-law Tom O'Malley, a captain in the PFD, and his wife Denise O'Malley.

RONALD R. SPADAFORA

Thirty-five years as a member of the FDNY. Do you think Ron Spadafora has stuff to pass on? Seriously. Ron currently holds the rank of assistant chief (3-star), and to give you an example of his experience, in 2005 he responded as the deputy incident commander with the FDNY IMT to New Orleans to assist the N.O. Fire Department in dealing with Katrina. On 9/11, he was recalled to Ground Zero and supervised the rescue efforts at the North Tower (WTC 1) collapse area and fire operations at WTC 7. In October 2001, he was designated the WTC Chief of Safety for the entire recovery operation. In addition to that, he has responded to and commanded countless fires and emergencies. Ron is a college professor and holds three college degrees, including a master's. Ron's pride shows when smiling, he reminds us that with more than 34,000 applicants for the FDNY test, he was hired in the first class of 121 ("cream of the crop") off the top of the eligibility list (7,847) on September 2, 1978. Testing hard, crawling hallways, getting his education, promoting up, and leading has provided Ron with plenty to pass on to us.

Chief Spadafora overseeing recovery operations at Ground Zero as site safety officer

Risk vs. Reward: 7 World Trade Center

TASK FORCE

It was an extremely stressful situation at best on West Street as more than a dozen chief officers were gathered together at approximately 1300 hours to provide reassessment and feedback to Assistant Chief Frank Fellini, the fire commander in charge of the west side area of the World Trade Center, concerning 7 World Trade Center. The collapse of the Twin Towers by 1030 hours had caused large pieces of structural steel to smash into parts of 7 World Trade, setting banks of floors on fire. Earlier that fateful day FDNY firefighters had entered the building to rescue occupants who were injured, disoriented, or unable to leave and to facilitate evacuation of nearly 4,000 people. Subsequently, it was verified by the fire safety director and deputy fire safety director that all occupants had been safely evacuated from the building. The examination of the building needed to be performed quickly yet strategically because time was of the essence. The major issue was whether the FDNY should or should not attempt to establish interior firefighting forces to extinguish the fires inside 7 World Trade. During initial firefighting efforts earlier in the day inside the building (after the collapse of Tower 1), attempts to extinguish incipient fires were hindered by low water pressure. Risk vs. reward: Should firefighters be ordered back in to save the structure from the ravages of fire?

RECONNAISSANCE

As I walked northward and turned the corner onto Vesey Street, I could see structural damage along the south and west walls of 7 World Trade. Most prominent was a large gash in the center of the south façade starting at about the 20th floor and moving upward toward the top of this high-rise office building.[1] A perimeter reconnaissance, starting along the west wall of the building and moving clockwise around it, provided valuable information concerning fire conditions. Flames could be seen on the exterior of the south and west walls of the building at multiple floors starting at about the 20th floor and continuing up for another 10.

Upon conclusion of my perimeter survey, winding up in front of the south entrance doors, entry into the high-rise proved difficult, as black billowing smoke was being emitted from an underground vault below the breached sidewalk located nearby on Vesey Street. The vault contained a set of four tanks holding up to 36,000 gallons of combustible diesel fuel for generators that served various tenants of the building. The lobby and staircases in the lower section of the building were examined for fire conditions. Caution was demonstrated so as to not go too far up into the structure because this was a recon mission conducted without the protection of sufficient personnel with charged hoselines. Escalators and elevators were inoperable or shut down by building maintenance. Smoke conditions on the floors I examined from the east staircase were variable although I did not observe fire.

The building was equipped with both a sprinkler and standpipe system. The primary water source for the sprinkler system, however, was the municipal water supply. These water mains had been damaged by the collapse of the Twin Towers. Additionally, each stairway had a standpipe, providing firefighters with a 2½-inch outlet and house hoseline (1½-inch in diameter) with a ¾-inch nozzle. The standpipe system was also fed from the municipal water supply.

DELIBERATION

Chief officers (me included) ordered to provide reconnaissance for 7 World Trade returned to West Street in a timely fashion to meet with Chief Fellini and discuss the building's condition. Several important points were identified:

- 7 World Trade had been damaged by debris falling into it from the collapses of the Twin Towers. There was no telling whether the structural stability of the building had been compromised. An in-depth analysis by a structural engineer, over a substantial period of time, would be required to gauge the degree of damage. Unfortunately, the FDNY did not have that luxury. Operating inside the building would be risky from a collapse potential perspective.

- The building had numerous fires of unknown magnitude on multiple floors. The location of these floors was not definitive. It was determined, however, that the fires were burning well up into the building. It would take a large commitment of manpower and resources over a substantial period of time to stage an attack on the upper floor fires. How long these fires had been burning for was not known. It was assumed that the fires started as a direct result of the collapse of the North Tower. The fact that these fires had been ablaze for several hours made it less likely that they could be extinguished readily over a short amount of time.

- There was no water immediately available for fighting the fires. Supply water would have to come from the north many city blocks distance from the building's fire department connections. An engine relay operation would take time to coordinate and establish. It was questionable whether adequate pressure and water flow at upper floor standpipe outlets could be provided to sustain an offensive interior attack on the fires.
- The necessary equipment (hose, standpipe kits, tools) required to conduct interior firefighting operations was not easily available. Additionally, there were not enough handie-talkies. Communications is a key component at any fire incident. A fire of this scale required all members to be equipped with radios. Coordination of strategy and tactics both inside and outside the building would be negatively impacted by the lack of handie-talkies.
- Working under these conditions would place FDNY members in unwarranted danger. I spoke to Chief Fellini directly on this matter although I was not sure he heard what I was saying over the din of feedback from the other chief officers who had gone to the building. There were pro and con suggestions about whether an offensive attack on the fire should be conducted. I told him that there was a good possibility that we would lose more firefighters if we decided to enter 7 World Trade to conduct firefighting operations.

IMPLEMENTATION

At approximately 1430 hours, it was decided to completely abandon 7 World Trade. Subsequently, another order was given to leave the area surrounding the building. This decision ended the ongoing rescue operations at 6 World Trade as well as on the pile of rubble of what remained of the North Tower. Firefighters and other first responders were withdrawn as the fires in 7 World Trade intensified and spread. At approximately 1720 hours, three hours after 7 World Trade was abandoned, the building experienced a catastrophic failure. No building like it (modern, steel-frame high-rise) had ever collapsed because of uncontrolled fires. There were no casualties related to the collapse.

AFTERMATH

During the early morning hours on the next day at Ground Zero, I ran into Chief Fellini again on my way to a new assignment at 90 West Street. He came up to me and said, "Ronnie, you were right about not going back into 7 World Trade." I looked at him and acknowledged the attaboy. I stated, "You were listening to me?" I couldn't believe that with all the commotion going on the previous day the chief was able to process my input and use it to make what I consider to be the most important operational decision of 9/11.

NOTE

1. World Trade Center 7 was a 47-story, high-rise office building with a trapezoid-shaped configuration. The linear dimensions were approximately 330 feet wide along its longest northern wall by 150 feet deep. The building was of Type 1-C, steel-frame construction. Vertical (columns) and horizontal (girders and beams) structural elements supported a composite concrete and steel (Q-deck) floor. It was built over a pre-existing electrical substation that spanned approximately half the site. It provided power to lower Manhattan. Transfer trusses and girders located between the fifth and seventh floors were designed to carry loads between non-concentric columns above.

Dave Statter

A Reporter Who Has Passed It On for Decades

Dave and I became friends in 1987 when those of us in Loudoun, Virginia, were trying to move forward from a rural fire and rescue service to a suburban service. Dave was then in the early days of his career as a TV reporter who covered the Washington, DC, area. Dave's broadcasting career started in 1972, but a decade later he was out of work one snowy winter day. Around 4 p.m. on January 13, 1982, he heard on a scanner about a plane crash near DC's 14th Street Bridge. Dave called the WTOP Radio newsroom and said, "Look, you don't know me, but there's been a plane crash at the 14th Street Bridge." It turned out to be the infamous Air Florida flight that crashed into the bridge after takeoff, killing 78 people. Dave found a hotel room near the crash, asked the people to use their phone, and started giving "live" reports as he watched the scene unfold below. WTOP immediately put him to work as a news and traffic reporter. This soon led to 25 years as a TV reporter, retiring in 2010. From his early days as a volunteer firefighter and medic, to a dispatcher in one of the busiest fire departments in the country, to his time as a reporter and now a blogger, Dave has been passing it on for years.

Dave Statter with his wife (radio and TV anchor) Hillary Howard

Soon, the Most Trusted Source in News May Be the Local Fire Department

I've come to dread turning on the TV or checking major news organizations websites, Facebook pages, and Twitter feeds during breaking news. I'm always wondering what major fact they will get wrong this time. This is not some anti-press media basher saying this. This is coming from someone who did almost four decades as a radio and TV reporter and who has a lot of experience handling live coverage of significant breaking news events.

It's not the small facts they are getting wrong. It's the reporting of big things that just never happened, such as an arrest in the Boston Marathon bombings of April, 2013, two days before one actually occurred, or reporting that a Coast Guard boat fired on a vessel in the Potomac River during a 9/11 anniversary visit to the Pentagon by President Obama.

As one former colleague put it, it's a rush by the mainstream news media to see who can get it wrong first. The most important job of a reporter used to be to confirm the facts *before* reporting them. Think back to how many times you've heard the phrase "unconfirmed reports" from a news anchor. Unconfirmed reports are just rumors.

Going on the air and online with these rumors is fueled by a desire to compete with the instant information provided by social media. But if the information from these news organizations continues to be no more reliable than what Joe Schmo posts on Facebook or Twitter, they will lose what makes them special in the first place. Despite what their promo announcements say, the mainstream media may soon no longer have the most trusted names in news.

That's where you come in. This is an opportunity for fire chiefs and fire departments all over the country to become that most trusted source of news and information, particularly at a time of crisis in a community.

The tools to do this are free and available to everyone. Many departments across the country are taking advantage of this opportunity. By using social media and the Internet, they are becoming a valued, trusted and instant source of important information.

These departments are no longer waiting for the local TV station or newspaper to show up to cover their events. They've learned that a television transmitter or printing press are no longer required to get your message out. You now can reach out directly to the public and the news media every day via Facebook, Twitter, YouTube, and your department's website.

Many departments are using these and other platforms to keep the public and reporters informed on the everyday emergencies and other activities firefighters are involved in. More and more, from small communities to large, people are seeing this as a primary source of information.

Those who are doing this have found that when it does hit the fan and the major crisis occurs, the citizens and reporters are turning to a fire department's Facebook page or Twitter account looking for details. We no longer gather as a family to sit in front of the TV to watch the news at 6:00. We get news, pictures, and videos all day long on our smartphones.

The key to success in this area is providing an early and frequent stream of confirmed information about a critical incident. It's the electronic way of holding the public's hand at a time when they most need it and providing key information and instructions on dealing with the crisis.

It also helps mitigate rumors and misinformation that will naturally surround such an event. For example, when CNN announced the arrest that hadn't occurred, the Boston Police Department used its Twitter feed to immediately and frequently point out that CNN's information was wrong. The Tweets from Boston Police helped that bad story die a quick death.

A lot of fire chiefs still avoid getting their department involved in social media. Yes, there are plenty of reasons to be cautious, but most of those can be dealt with by having good policies, procedures, and training.

As we all know, modern fire departments protect their communities in ways beyond sending fire trucks and ambulances on 9-1-1 calls. Reaching the public with crucial information can also help save lives and avoid panic. If you aren't on Facebook and Twitter, you may not be reaching the public. If you are, you have the opportunity to be your community's most trusted name in news. Use it wisely.

Phil Stittleburg

Volunteers, Leadership, and "the Payday"

I have known Phil for years but have gotten to know him closer as we both serve on the board of the National Fallen Firefighters Foundation. He also serves on the National Volunteer Fire Council as the Director from Wisconsin since 1979, and has served as Chairman of the NVFC since 2001. Phil has spent 36 years spent prosecuting bad guys as assistant district attorney in Vernon County, Wisconsin. He retired in 2010 and now runs a private law practice in LaFarge, Wisconsin. He also serves as chief of the LaFarge fire department, taking care of 3,000 people across 135 square miles. He's done that for four decades, chief for much of that time. Making the tough decisions, be it in the courtroom or at the fireground, seems to be a personality trait. He tells people that what really motivates volunteers is payday. "Payday?" When you work for free, "thank you" is payday. When you cut somebody's kid out of a crash, and the parent says, "Thank you for saving my kid's life." Or when you save someone's house, and you hear, "Thank you for saving every treasure I have here on Earth." You can really work hard for a long time for that kind of payday. Fortunately we get a little real payday now as well as Phil passes it on.

Phil Stittleburg

A Thousand Thank Yous

The volunteer fire service is an amazing institution, and being a member of it is a great honor. Volunteer firefighters carry on a wonderful tradition and are in very distinguished company. Over the years, the rolls of volunteer firefighters have included the likes of George Washington, Thomas Jefferson, Samuel Adams, John Hancock, and Paul Revere. This amazing tradition has been passed down from generation to generation and finally passed to us. What an honor to be able to advance the torch and pass it to those who will follow.

To continue this tradition, we must understand what motivates volunteers. Personally, I think that payday is what motivates them. Obviously, it's not payday in the conventional sense. However, when one works for no pay, "thank you" is payday.

It is important for us to recognize "thank you." Sometimes it's obvious. When someone comes up to you and says, "Thank you for putting out my house fire and saving every possession that's important to me in this world," or, "Thank you for cutting my child out of that crashed car and saving his life," that's easy to recognize.

More often, though, the public doesn't say "thank you" in so many words, but it shows it and we need to learn to listen. When the public supports our fundraisers, when the governing body approves our budget even though money is scarce, when there is a note of pride in a family member's voice when they introduce us as firefighters, that's "thank you." We've just been paid.

The volunteer fire service recruits from every walk of life. Every profession, every occupation, every religious preference, every political conviction, is represented and appreciated. Membership cuts across social and financial distinctions and is truly democracy and the American melting pot in action. It unites people who have nothing more in common than the desire to protect the lives and property of their neighbors.

For volunteer fire departments to survive, that desire must be very strong. It must be stronger than personal preference and stronger than the personalities of the individual members. We require people who are dedicated, intelligent, ambitious, and civic-minded.

Of course, there are challenges on the horizon. There always have been. The two biggest are time and money. Believe it or not, money is probably the lesser of the two challenges. The time has come to encourage local and state governments, as well as the federal government, to pick up more of the cost of operation. As the public demands more and better services, there is less time for volunteer firefighters to spend raising funds. Although this may mean involvement in the political process, it does not mean partisanship. We simply need to become more effective at making our case to our elected officials.

Time is the more difficult challenge because it is the finite resource. In other words, given enough time, we can make more money. However, given enough time, we can't make more time. We can hold a fundraiser, but not a time raiser.

Always keep in mind that our members are our most important resource and that their time is their most important resource. We owe it to them to use it wisely. If we squander their time, then we squander their most valuable contribution—and worse yet, we send them the message that their contribution is unimportant to us. Be efficient in training.

Time available for volunteering is becoming scarcer as the pace of life continues to accelerate. This means that we need to change our recruiting mindset. Traditionally, we have required people who contribute to the fire department to first be a firefighter. Then we expect them to take on a number of nonoperational duties, such as public education, maintenance, record keeping, and fundraising, just to name a few.

This approach immediately disqualifies people with disabilities, those with prohibitive time schedules, and those who simply do not want the inconvenience or danger of firefighting, but would like to make some contribution to their local volunteer fire department. Encouraging the public to assist their volunteer fire department in nonoperational roles is one solution. The retired teacher may be interested in assisting with public education. The local CPA may help with budgeting. The local mechanic may lend a hand with maintenance. With every hour covered in this fashion, a corresponding hour is restored to a volunteer's schedule.

Undoubtedly, the future will continue to bring challenges. How we meet these challenges depends upon how we view them. Always bear in mind that regardless of whether we believe that we will succeed or are convinced that we will fail, we're right. I believe that we will succeed. We are problem solvers by nature. Every time the pager goes off, someone has a problem for us to solve. We have an unbroken 300-year record of success in solving problems and meeting challenges. I predict that we will continue this tradition.

John B. Tippett, Jr.

Positive Results from Tragic Circumstances

I got to know John when he was involved from the start with the National Fire Fighter Near-Miss Reporting System. He was a fire battalion chief in Montgomery County, Maryland. What stood out to me was his sharp focus on the mission, and getting results. On June 18, 2007, the unthinkable happened at a Charleston, South Carolina, furniture store fire that killed nine firefighters. John's chief at the time in Maryland, Tommy Carr, ended up taking command of the Charleston Fire Department. His challenges were significant not only to help a department heal, but significantly change every aspect of its operations. To help accomplish that, in 2009 Tommy recruited John for the deputy chief position. A different form of sadness again struck the department when Tommy passed away at 59 from complications from Multiple System Atrophy, a Parkinsonism disease. Again John found himself in a position to stay focused on what Tommy had wanted and what was needed in Charleston.

John B. Tippett, Sr., John B. Tippett, Jr., and Tommy Tippett

Keep Your Head When Others Are Losing Theirs

It was the first weekend in December, 1968. I was 10 years old and "helping" my dad put up Christmas lights on our house. He had put a ladder up against the gutter (three rungs over the roof line), set it in the soft earth, and climbed up on the roof with several strings of perfectly coiled lights slung over his shoulder. He was carefully uncoiling the rolled lights across the roof in preparation to attach them to the gutter, taking care not to drop them and smash the delicate green and red painted glass. There was a methodical approach to his work that was without equal. I had watched other dads hang their lights weeks before. They stood in their front yards looking as if they were wrestling a multicolored, many-tentacled octopus; but not my dad. His lights were always neatly coiled and carefully uncoiled, even as he was gently laying them out on the roof.

This was going to be the year I got to really "help." I was old enough to climb the ladder, stand at the tip, and watch my dad put the lights on the house. I thought it was cool to be so high off the ground, and I watched my dad with pride and awe as he moved nimbly on the roof. You see, he was very comfortable working from heights. He was a strapping man, athletic and well coordinated. He played semi-pro football and was trained to climb telephone poles as part of his service in the Signal Corps. He was proud to have served his country in the Korean War. That training and service led to a career with the telephone company. Although he would eventually rise to first level supervisor, at this Christmas season, he was still an installer. Most days he was up and down telephone poles, running lines across streets and alleys, and connecting people.

The annual tradition—Christmas activity wasn't allowed to start until after Thanksgiving—was well under way when one of my most significant life lessons for my future career occurred. My dad was about 25 feet away on the roof and I was at the top of the ladder. I was testing my reach when the damned thing started to slide along the gutter, away from my dad. The slide was slow at first, but then started to pick up speed, heading toward the tall Hemlock tree that anchored the corner of our house. I gripped the rung, fear rising from my stomach, and looked at my dad, who had just looked up from laying out the last string of lights on the roof. Without hesitation, he started toward me at a dead run across the roof. Before I knew it, he caught up to me, grabbed the beam closest to him, and stopped the slide instantly with one hand. I can still remember my heart pounding and looking at him. Without skipping a beat, or even being out of breath, he quietly said, "Where ya goin' boy?" He pulled the ladder back to center and had me come up on the roof with him. He never said another word about the incident, and we went on to finish the lights as if nothing happened. After we finished, we climbed down from the roof; he descended first and positioned himself to support me as I completed my first descent of a roof via ladder.

Forty-five years later, the lessons that event taught me, and the way my dad lived his life, have grounded me throughout my life. The event was a microcosm of how to serve others. Be organized and methodical. Stay calm when everyone else's hair is on fire. Keep your fears under control and to yourself. Help others. Do the right thing, don't make a big deal out of it, and then go back to work. On a grander scale, be thankful for your job, loyal to your friends and coworkers, live with discipline, don't burden others with your problems, and take care of others before yourself.

Those lessons, reinforced by a long line of other good men and women throughout my career, were passed from my father to me, and hopefully from me to my kids. Those powerful, yet understated lessons have been invaluable. The effect of remaining calm under pressure, being determined and focused, patient and persevering, conducting oneself with a stoic demeanor, and keeping others' needs ahead of your own have been immeasurably valuable as I transitioned from company officer to battalion chief to deputy chief. The collateral assignments I've had as safety officer, instructor, document control officer, task force leader, and most recently operations chief were proving grounds in practical application. Each role played out on a stage of constant change. The latest role, operations chief for a department forced to make exponential change on a very restrictive timeline, has required the greatest application of my father's example.

Even though my dad wasn't a firefighter, his approach to life was the perfect preparation for doing a firefighter's and officer's jobs. Now that one of my own sons has entered the job, I was hoping I had done an adequate job passing those lessons to him. In conversation with my son while he was in recruit school, and since his graduation, something seems to have taken hold. One of the earliest experiences my son relayed with enthusiasm after graduation gives me confidence. It was not a fire or another adrenaline-rush event. It was an account of how his captain had kept a promise to a little boy. The company my son is assigned to met a mother and her young son in the supermarket while picking up groceries for the company meal. After talking with the child and his mother, the captain promised to bring the engine by the little boy's house for a visit on their next tour.

The captain made good on his promise. When my son's company arrived at the little boy's house, they discovered the boy is seriously ill and his brother is a child with special needs. My son relayed how rewarding it was to see the little boys' reactions and humbling to hear the parents' appreciation for so simple an act. He said the interaction with the children and his crewmembers felt more fulfilling than putting out a fire. My son mentioned the deep respect he had for his captain for keeping the promise he made to the little boy. As I listened to my son, it seemed to me as if those lessons of my father, and those of other quiet but strong mentors I have known through the years, had been passed to another generation. The job is not about what's in it for me; it's always about serving others.

Matt Tobia

Unbridled Energy to Do the Right Thing

I got to know Matt years ago when we taught firefighter survival training in the early 90s together. It didn't take me more than about 30 seconds to understand that this young man was kinda like a fire service's Tasmanian devil . . . after about 50 cups of coffee. Matt gets wound up—and he should. Too many people in our business don't, so Matt has it covered. What does he get wound up about? When one of us gets critically hurt or killed—for reasons that could have easily been avoided.

He's been there—he knows.

Matt works directly with the families of fallen firefighters, serving as the Family Escort coordinator at the annual National Fallen Firefighters Memorial Weekend in Emmitsburg, MD. In addition, he is a counselor at the Mid-Atlantic Burn Camp for children.

Matt is a battalion chief with the Anne Arundel County, MD, FD. He holds a bachelor of science degree from the University of Maryland and is a nationally certified Fire Officer IV and Instructor III. He is the chair of the IAFC Safety, Health, and Survival Section, and he teaches extensively throughout the United States, lecturing frequently on safety and leadership issues. He writes the monthly "back page" column in *FireRescue* magazine. And Matt remains wound up. His energy includes consistently passing it on—so we don't repeat history, history of which he has seen first hand from the families left behind.

Service above Self

This is dedicated to my wife, Jeanne, who I love with all my heart, and to the families of our fallen firefighters—we will never forget.

The fire service does not belong to us. It was here long before any of us, and it will be here long after we are gone, and it will endure. That is as it should be. We are stewards of a gift to which we have been entrusted, and we have been called upon to protect its reputation with every fiber of our being. The fire service will never know perfection, but our job, *your* job, is to leave the service better than we received it. In order to do so, you must understand the depth of the promise you have made in becoming a firefighter. From this day forward, and as long as you are a firefighter, you must conduct yourself in a way that is different from any other person. You will be held to a higher standard by citizens who demand greatness. It is not for us to determine the validity of such expectations. Such is the price of having the honor of being able to call yourself a firefighter.

Your first commitment is to your family—not only to their well-being but to ensuring that you come home at the end of every alarm. Honor comes not from dying in the line of duty, but *living* in the line of duty. The only circumstance when firefighters should be prepared to sacrifice their lives is in the act of placing themselves between harm and those we serve. Self-indulgent, ego-driven actions may be the stuff of legend, but they do nothing to protect the reputation of our calling and betray our primary responsibility.

Your next commitment is to those firefighters who cannot speak for themselves—our fallen brothers. Attend the national memorial service at the National Fallen Firefighters Memorial in Emmitsburg, MD, meet the families, and pay respect to those who have made the ultimate sacrifice. Then go out and do everything in your power to prevent similar events from happening. The single most disrespectful thing you can do to the memory of a fallen firefighter is to ignore the lessons to be learned from their sacrifice. It is as bad as killing them all over again.

Leadership is born from followership, so choose carefully who to follow, but do not follow blindly, as this can lead to tremendous disappointment when you reach the conclusion that no individual is perfect. This does not mean that those you aspire to model cannot offer you tremendous lessons to live by . . . only that they are human. Judge for yourself and think for yourself. Following individuals who, through their actions, demonstrate the essential qualities of integrity, honesty, and courage can yield a lifetime of benefits, the magnitude of which cannot be measured in words alone.

You will never be able to pay such individuals back for their commitment. You must therefore pay forward by offering to those coming after you the same opportunities. This is the essence of keeping the fire service alive. Live your life in a way that brings credit to those who made your life possible and when you receive compliments, give credit to those who so aptly deserve it: your parents and those who showed you "the way."

There is no substitute for competency. You may be the nicest person in the world but if you are not competent, firefighters will not trust you with their lives. Competence is the natural product of an internal demand for perfection. It is the product of relentless preparation through training, education, and experience. You must be physically fit, mentally and emotionally healthy, and do everything in your power to lead a balanced life. Never forget that a "good fire" means someone we have sworn an oath to protect is having the worst day of their life. You may lessen, but will never erase, the pain of such events; always remember that humility is the cornerstone of our service.

Theodore Roosevelt, one of our greatest leaders once said:

> It is not the critic who counts; not the man who points out how the strong man stumbles, or where the doer of deeds could have done them better. The credit belongs to the man who is actually in the arena, whose face is marred by dust and sweat and blood; who strives valiantly; who errs, who comes short again and again, because there is no effort without error and shortcoming; but who does actually strive to do the deeds; who knows great enthusiasms, the great devotions; who spends himself in a worthy cause; who at the best knows in the end the triumph of high achievement, and who at the worst, if he fails, at least fails while daring greatly, so that his place shall never be with those cold and timid souls who neither know victory nor defeat.

> Honor comes not in the title firefighter, but in your actions as a firefighter.

Arlington National Cemetery at the Tomb of the Unknown, my favorite place to visit in Washington DC, completely embodies the concept of "service above self."

Parker Crow, grandson of Assistant Chief Brian Hauk of the Logan-Trivoli Fire Protection District (LODD 12/23/1997), on his first visit to his grandfather's gravesite.

Bruce H. Varner

An Interesting Firefighter, Developing the Standards for Our Future

Like in the commercial, Bruce is one of those "most interesting" men. Starting in 1967 as a firefighter in Phoenix, Arizona, he moved up to deputy chief in 1985 during an interesting period when Alan Brunacini was making sweeping operations changes. After Phoenix, he served as chief in Carrollton, Texas, and then on to Santa Rosa, California, where some of the best omelets were ever made. In his career, Bruce has made an impact on every firefighter in the U.S. and Canada well before "firefighter safety and health" became a commonplace idea. He was often the lone voice in the room pushing the issues most of you understand today. From intelligent aggressive firefighting operations; to command, control, and accountability; to firefighter survival; to tactical fire scene leadership and building construction; as well as SCBA and related technological development and standard establishment, Bruce has been there from *day one*, pushing our service forward, sometimes kicking and screaming along the way. Involved with the NFPA process for decades, Bruce looks out for us and has a few things to pass on.

Bruce Varner at Santa Rosa Fire Department

45 Years and Still Learning

September 18, 1967, was my first day as a recruit firefighter in Phoenix, Arizona. It was a three-week academy, just recently extended so we could complete an American Red Cross Basic First Aid course before being assigned to a fire company. "Do it the right way every time" was the mantra of the day. We stretched hoselines, threw ladders, chopped on old telephone poles with axes, tied knots, and pulled more 2½-in. and 1½-in. hoselines. Safety was voiced as "pay attention, listen to your officer, and do it right every time"

We were also shown the two SCBAs carried in the boxes, kept in the basket over the hosebed on each engine, and taught about the Type N filter canister mask, which was the breathing protection of choice. Only wimps wore SCBA, unless it was a basement fire. During the last week of training, an explosion occurred while companies were operating at a scrap metal yard fire; magnesium shavings were the culprit. A third alarm was stuck; several firefighters were injured, including one with serious facial burns. As our class of nine recruits rolled and picked up hose at the scene we heard stories about what had happened: "just bad luck," "could happen to any of us," "no one knew there was magnesium in there." Fortunately everyone recovered and returned to work over the next couple of months.

No seat belts, no safety straps, no buzzers, "hang on tight and don't fall off." I had the good fortune of being assigned to Captain Kenny Simmons at Engine 15 as my first station. Captain Simmons believed that we could and should be doing things more safely. He talked about staying with him, not getting off the rig until given an order, paying attention, and doing it right the first time. He was always the last one on the rig, making sure that his tailboard firefighters were ready before telling the driver/engineer to proceed. We talked about every response when we returned to the station. We drilled, practiced, and discussed expectations, performance, and what the public expected from us. We operated in a mixed, commercial, industrial, and residential response area, which gave us some really good opportunities to learn our craft in our customer's buildings. It was a great start to a 45-year carrier.

The Phoenix Fire Department became one of the finest and safest fire departments in the United States thanks to the leadership of an amazing and truly wonderful fire chief, Alan V. Brunacini. He changed the organization and he changed the fire service in the United States. His book *Fire Command* (still the simplest most effective system for managing Type IV and V incidents) described the operating principles that he developed with the shift commanders and executive staff of the department. We became close friends; I was privileged and fortunate to be a part of the "Fire Command Seminar" program and spent 15 years traveling to fire departments all over the country delivering that program with Chief Brunacini. It was a fantastic opportunity to learn, discuss, listen, and be mentored. I firmly believe that experience was one that helped me to become a fire chief later on. Along the way I developed a passion for firefighter safety, became involved in the NFPA Protective Clothing and Equipment project, and helped set standards for firefighters.

There have been a number of changes in our profession over the years; we have seen improvements in fire apparatus, protective clothing, respiratory protection, communications capabilities, fire hose, the development of compressed air foam, health and fitness capabilities, and in the not-so-distant future will have the ability to track firefighters inside structures while monitoring their physiological status. We will see small rapidly deployable unmanned aerial systems capable of giving the fireground commander a quick look at all sides of the building or incident scene, utilizing both streaming video and thermal imaging.

As a service we have been quick to embrace many of these improvements and very slow to adopt some of the others. *Why:*

Are we still having firefighters killed and injured because they are not wearing their seatbelts?

Are we still seeing respiratory and smoke inhalation injuries?

Are firefighters still injured and killed inside well-identified collapse zones?

Do firefighters still get lost on the fireground, separated from their other crewmembers?

Are firefighters killed and injured in backing and tanker/tender rollover accidents?

Do firefighters refuse to get an annual physical or take best-practice basic medical and cancer screenings?

Are we still seeing deaths from previously identified cardiac conditions?

I am sure that I missed something in the list, but if we just addressed all of these, would we be meeting the National Fallen Firefighter's Foundation (NFFF) firefighter death reduction goals? We have made progress, but we can do better, much better!

Now, let's talk about fire behavior for a moment. The fire has changed. Really? I would wager that fire still burns the same as it always has; fire is a physical process. The building construction has changed, the stuff that we a putting in those buildings has changed. We are, in fact, seeing fires that progress much more rapidly than our predecessors experienced. National Institute of Standards and Technology (NIST) and Underwriters Labs (UL) have documented fire behavior and a need to change or adapt our tactical approach to these fires. There is more material available today on this topic that ever before in our history. Some in our profession have taken the information and are adapting, whereas others persist in resisting the scientific facts. Is it because we still think of ourselves as Knights of the Realm and want to go "dance with the dragon?" Is it because we are smarter than physics? As one very wise sage said, "Put water on the fire, as fast as you can, from as far away as you can." And live to fight another day. Take the time to review the available material, use it for training, use it for decision making, use it to keep your crew and yourself safe.

Ask yourself, would you want your son or daughter to be a firefighter? For almost any of us the answer is of course, it is quite simply the greatest job anyone can have. Would you want them to survive the experience? Again the answer is simple, so let's all work toward making the changes that move our profession in a much safer direction.

CURT VARONE

Not Your Typical Firehouse Lawyer: This One's Actually Qualified

Several years ago, I met Curt when we taught for the old Command School group, run by Chief Glenn Usdin. We traveled the country doing seminars primarily on firefighter survival. I met Curt and was told he was an attorney and a fire chief. It confused me until I found out he was one of the good ones. (Chief or attorney? Possibly both.) Curt retired from the Providence Fire Department in 2008 as a deputy assistant chief and served as the director of public fire protection at the National Fire Protection Association. A double major college graduate and cum laude graduate of Suffolk University Law School, Curt has engaged in the general practice of law. He teaches at FDIC, Firehouse, and other venues, and runs the Fire Law blog. As an urban firefighter as well as a volunteer firefighter in the suburbs, he has a lot to pass on, especially when it comes to the worst experiences of those runs that impact the public—as well as the brothers and sisters responding.

Curt Varone

Give Me Five Minutes, Kid; Sit Down and Listen to What I Want You to Know from My View of the Fire World

It's not about you.

Being a firefighter is a privilege that not everyone has the opportunity to enjoy . . . nor even realizes is a privilege. It is about having the opportunity to help people who are in need—and if there is one thing I can impress upon you, it is that it needs to stay about helping people. Don't let it become about you.

Just as many citizens and politicians take firefighters for granted; many firefighters take being a firefighter granted. Cherish every minute you have in the fire service. Make the most of every opportunity you are given because in doing so you are helping to change the world for the better.

Firefighters are the greatest people in the world. They are the most selfless, generous, and caring people imaginable. If Jesus Christ were alive today he would be a firefighter. He would not be giving holier-than-thou sermons in a church or going on television asking for money. He would be out there on the front lines doing the Lord's work with His own two hands. His sermon would be the way He lives His life.

At the heart of being a firefighter is the satisfaction in knowing that what we do matters . . . it makes a difference. As firefighters, we are our brother's keeper. Even when there are no alarms, just being available to help people matters. It gives meaning to life beyond most other endeavors we could do on this earth. Being able to help people in this way is the ultimate expression of caring . . . of love . . . and in this regard truly is a privilege.

The desire to help and serve can be a source of motivation and inspiration throughout your career. Let it guide you to keep improving yourself to be the best you can be.

Throughout your career there will always be reasons . . . excuses . . . for doing less than your absolute best. Someone will criticize you, not appreciate you, or take your efforts for granted. It is easy to become indignant when our services and even our sacrifices go unappreciated. It is easy to become self-righteous when our efforts are trivialized or scorned by those who rarely lift a finger to help anyone. But becoming indignant and self-righteous means we have allowed it to be about us, not about the people we serve. Our service cannot be about us. It has to remain about our service to others in their time of need.

Today many elected officials find it politically advantageous to criticize public employees because they believe that is what will get them more votes from a public that is fed up with the dysfunctions of government. If being a firefighter is about us—if it is about being recognized, appreciated, or even treated fairly—then it is easy to fall into a cynical, self-righteous trap. There will always be a justification for a lousy attitude.

There are career firefighters who rationalize why it is permissible to abuse sick leave, somehow justifying in their own minds why committing fraud is acceptable. There are volunteer firefighters who rationalize why it is appropriate to use fire department funds to provide free beer in the station. There are career and volunteer firefighters who rationalize reckless behavior on the fireground as well as while driving apparatus. This list of rationalizations could go on and on.

The world is full of selfish and self-dealing people who look to rationalize why they should be allowed to do whatever it is they decide they want to do. Unfortunately good people sometimes fall into the rationalization trap as well, seemingly pushed there by the forces of destiny.

My advice to you is to rise above the selfish rationalizations and justifications that so often confront us. The way to do that is to keep it about the people we serve. Make sure your role in the fire service is grounded in your service to others and is not about "what's in it for me." It's not about you. When it becomes about you, self-righteousness becomes inevitable. Avoid the temptation to rationalize why you are entitled to something. Stay off that path. Focus on service to others.

Remember: It's not about you.

Reflections on The Station Night Club Fire

It started out like any other night shift. I had worked my normal A-Group day shift in Providence (08:00 to 18:00), and because the deputy chief for D-Group was off, I stayed to work the night shift (18:00 to 08:00) as well.

The evening was uneventful, mostly spent organizing personnel to work overtime for the following day shift to meet the minimum staffing requirements. I recall lying down at about 11:00 p.m. hoping to catch some shut-eye, mindful that we rarely had a "good night" at the Washington Street station. The companies assigned to the station responded to over 13,000 runs annually and it was unusual for more than 45 minutes to go by without the alert tones sounding. Given that I had to work my normal shift the following morning, sleep was an imperative.

Shortly after I dozed off, the department telephone rang. It was our dispatch office notifying me that we were sending companies on mutual aid to fire in West Warwick. It was a courtesy call, the kind that the shift commander in Providence routinely receives several times a night. After some small talk with the dispatcher, I resumed my battle to fall asleep.

A few minutes later the dispatcher called again to notify me that we were sending some additional companies out of town to cover some stations in the city of Cranston because Cranston had dispatched a large number of trucks to West Warwick. When I asked what was going on, the dispatcher said there was a restaurant fire on Cowesett Avenue. I asked if there was much to it, and he said he did not know, but would contact me if he found anything else out.

Given the late hour, my assumption was a restaurant would have been closed at the time, and I didn't think much more about the fire. We discussed the distribution of companies across the city of Providence and, once assured we were adequately covered, I resumed trying to get some sleep.

As the evening wore on I caught a couple of runs but was cancelled en route to each by the first-arriving companies. Each time, it was back to the station where I would again try to doze off. As best I can recall it was about 3:00 a.m. when dispatch called again with a rather strange request. "Chief, can you give the fire chief from West Warwick a call on his cellphone?" The dispatcher started to give me his number but I did not pay attention. The chief was a personal friend whose number was already in my cell phone.

"Chief—it's Curt, what's up."

"Chief, I need a favor," he said.

"What do you need?"

"I need you to send some guys over to the medical examiner's office to help."

"Help with what?" I naively asked. He assumed I knew about the fire. I was thinking he had an after-hours fire in a restaurant.

"We had a very bad fire—there are a number of bodies, and the medical examiner's office is going to need some help unloading them."

"Firefighter fatalities?" I asked hesitantly with the faces of my friends who are on the job in West Warwick flashing through my mind.

"No . . . no . . . civilians. It was a nightclub fire and we have quite a few fatalities. I think you'll need eight to ten guys, so maybe two to three companies if you can spare them."

"No problem. When do we need to be there?"

He said, "Plan on about an hour from now."

After we hung up, a flood of thoughts went through my mind: Did I know any of the victims? Were any of them children? Nah . . . it's a nightclub. But some may not be much older than my own teenagers. Maybe even some of the boys I coached in baseball. Where is there a nightclub on Cowesett Avenue in West Warwick?

My mind reflected back to a recent discussion I had with a training academy classmate of mine over whether firefighters should be responsible for handling deceased bodies. A year earlier we had a five-fatality fire and several of our members who carried out the gruesome task of placing the bodies in the body bags were still troubled by it. One firefighter had not been able to return to work.

At the time, I posed the question about whether handling bodies in nonrescue situations should be our job, suggesting that perhaps going to fires with the hopes of saving a life, and then having to carry out someone's remains might be more painful than most firefighters can humanly bear. My friend was incredulous that I would suggest such a thing. He argued that of course it was our job, and why would it be easier for a morgue worker? It was one of those conversations where I was thinking out loud, struggling to find a solution to a dilemma that had no easy answer.

Now I was about to order firefighters to handle another gruesome task involving bodies. I shook my head as I realized that my friend would soon be reminding me of our conversation, perhaps even calling me a hypocrite.

I went to kitchen and grabbed a cup of coffee. It was 03:15, and any hope for sleep that night was over. Looking into the sitting room I saw a half-dozen firefighters glued to TV as the news kept replaying a video clip of the fire starting. The news was reporting at least seven dead, perhaps as many as ten.

Back in my office, I ran through a list of companies I had available, checking the names on the rosters to determine who was working. I wanted to avoid companies with rookies. I also wanted companies with strong, experienced officers. I was fortunate that many of my A Group officers were working overtime that night. While I knew and respected virtually all of the D Group officers, I knew my own officers better. This was going to be no easy task. I selected three companies, two engines and a ladder, which I felt would be up to the assignment.

I spoke personally with the officers of each unit via the department telephone and explained the situation. I instructed them to meet me at the medical examiner's office at 04:00.

We arrived at the medical examiner's office to find one person working. He thanked us profusely for coming and showed us around. The smell of death was rather pungent in the air, so after the brief tour most of us chose to wait outside on the loading dock. It was cold, bitter cold, but the cold was preferable to remaining inside.

As we were waiting on the loading dock, two men came walking rather sheepishly around the side of the building. They appeared to be in their mid-60s, and I would have taken them to be retired police officers, or perhaps retired firefighters. One was wearing a U.S. Navy baseball cap with a ship's identification number on it. For the hour they seemed remarkably well attired and wide awake.

I asked, "Can we help you?"

The man without the cap said almost apologetically that he was looking for his daughter. Stupidly and again naively I asked, "Does she work here?"

"No—she was at the nightclub." In an instant, the gravity of what had occurred came crashing down on me.

"We have been to all the area hospitals and she is not there." As he stood there clear-eyed, I recall being amazed at his calm demeanor. No tears, no quivering in his voice. It was as if he were asking if his dry cleaning was available for pickup, not whether his precious loving daughter was among the dead from a tragic fire.

After what seemed like an eternity, a caravan of vehicles arrived with a police escort. The first vehicle was the medical examiner's suburban. It contained two bodies, both in thick, well-constructed brown colored body bags with sturdy zippers and strong handles. Next was a funeral parlor hearse driven by a former firefighter I knew from North Providence. David's eyes said it all, the classic 1,000-yard stare.

After we unloaded the two bodies from David's hearse, I glanced around the corner of the building expecting to see two or three more vehicles. There were at least another dozen. The next vehicle had backed in, and it had three bodies. The next was a pickup truck with five bodies.

After moving his vehicle off to the side, David reappeared, climbing up on the loading dock to help with the lifting. I asked him, "How many more vehicles are there?"

"This round?" he replied. His response floored me.

"What do you mean, I thought there were seven to ten fatalities. We have at least twenty bodies here now."

"Chief, this is just the beginning."

The loading dock was too high for the most of the vehicles that arrived to back up to. As a result we had to position firefighters on the ground to remove the body bags from the vehicles and then lift them high enough for those on the loading dock to reach. The firefighters on the loading dock would then lift and carry the bags into the building.

We carried the bodies into the available rooms. The first few were placed on carts, but it soon became apparent that the number of bodies vastly outnumbered the available carts. The morgue was not at all like what you see on television. Bleak concrete walls and floors. No shiny stainless steel counters, no fancy roll-out drawers for the bodies, and no fashion-model CSI technicians. The smell of death and burned flesh was overwhelming.

After the first round of bodies had been moved inside, we were informed that it would be a few hours before the second round would arrive. The morgue technician asked us if we could assist him with the identification and tracking of the thirty or so bodies. With the same tenderness and respect given to conscious patients, firefighters unzipped each bag, identified the victim as male or female, estimated the approximate age, looked for any distinguishing features or marks, and determined if they had any identification in their possession.

The firefighters performed this task without hesitation, grumbling, or commentary. To this day I am humbled and incredibly proud of their service. Had any of the victims' families been there to watch they would undoubtedly have been touched by compassion displayed by the men. It helped also that Fire Chief Jim Rattigan and Assistant Chief Gary Mulcahy arrived and assisted. They literally rolled up their sleeves and, perhaps forgetting their position and rank, worked side-by-side with the firefighters in carrying out the gruesome assignment.

We had the opportunity to return to quarters between the first and second round. I am not sure if that was a good thing or a bad thing. The thousand-yard stare that was evident on David's face was no doubt evident on ours as well.

There would be three rounds with which Providence firefighters were tasked with helping at the medical examiner's office, the last occurring just before noon. There are so many recollections and emotions from that morning . . . sights, sounds, and smells.

Afterward, every Providence firefighter involved in the operation voiced a very similar sentiment: as bad as what we experienced was, it paled in comparison to what the firefighters at the scene had to go through. And it goes without saying that what all of the responders experienced in responding to The Station Nightclub fire pales in comparison to the loss suffered by the families of the victims.

Probably the most significant recollection that stands out in my mind was the marked difference between the condition of the bodies and the quality of the body bags between the first and the third rounds.

The bodies of the victims in the first round were largely intact, and the bags were high-quality thick plastic with substantial zippers and strong handles. By the third round most of the victims were very badly injured and the bags were an assortment of colors, some brown, some black, some green, but most were white. It wasn't so much the color of the bags that stuck with me. What remains etched in my mind is that by the third round the bags were not as thick, the zippers were not as substantial, and most importantly the white bags lacked handles.

You may ask why the absence of handles would matter that much. My sincerest hope is that you never have occasion to find out. What I can tell you is that as the firefighters would return to the loading dock after carrying a victim, we each hoped our next carry would not be a white bag.

The Station Night Club fire changed the lives of thousands of people. If affected each of the firefighters who worked at the medical examiner's office in different ways. Some left the service afterwards, unable to cope with the horrors they witnessed. It undoubtedly changed the way I look at the world. My changed perspective can best be summarized in the following way.

Only someone who has never had to carry a body bag would consider purchasing body bags that are so cheap they do not have handles. That may seem like a harsh overstatement, but in my mind it is an inescapable truth. And it doesn't stop there.

The same could be said for the owners of nightclubs who choose not to install sprinklers, of city and state officials who choose not to hire enough fire inspectors, and politicians who choose not to provide enough firefighters to properly staff fire apparatus.

Only someone who has never had to carry a body bag would consider not installing sprinklers in a nightclub. Only someone who has never had to carry a body bag would consider not to hiring enough fire inspectors to properly conduct inspections. Only someone who has never had to carry a body bag would consider not providing enough firefighters to safely staff fire trucks.

Perhaps there is a solution to my dilemma after all.

Colleen Walz

Leading Justifiable Anger to Measurable Action

I met Colleen following a line-of-duty death fire. She had been involved with other LODDs in her career, and what struck me about her was how angry she was, specifically with respect to the issue of preventability. She never forgot those losses, and none of us would. She never accepted the excuses, barriers, and traditions, insisting that the truth be shared, that lessons learned would be the best way to honor those who were killed. Colleen is now the fire chief of the St. Johns Fire District, Johns Island, South Carolina. Prior to that, she spent more than 25 years in the Pittsburgh, Pennsylvania, Bureau of Fire. Among her many accomplishments, she co-chaired the IAFC's Safety and Health Section, Firefighter Fatality Investigation and Prevention Program Task Force, outlining significant recommendations for NIOSH. She's an EFO graduate, has her master's degree in leadership, is a college instructor, and, right now, here are some heartfelt words to pass on to us.

Colleen Walz

It's Not a Cliché

You never get over the death of a firefighter in your department. Maybe I should say most of us—those who will admit it—are forever changed when a traumatic death of a firefighter occurs. If you aren't changed, you are lying to yourself. Publicly most times you only hear from the chief, who may or may not have originally come from the department. That fact doesn't make the situation any less tragic, but what does that death mean to the firefighters left who have lived with and bunked with that firefighter? The firefighters who went through recruit training with that firefighter? Who ran calls with that firefighter, and who had your back so many times before?

Often you hear from the big chief in charge of the department that under his or her watch this tragedy occurred, but you—the firefighters who worked side by side—can taste the heartfelt grief of the department as you try to swallow that lump in your throat when you see that grief on the faces of the families. Ironically, just as hard to swallow are the comments from the rank and file shortly thereafter, the familiar, "it's what we do," and "those were the risks we knew about when we took the job" mantras we who have experienced line of duty deaths often vocalize over and over again. I think those mantras are a way we try to rationalize or convince ourselves that what will be will be, or somehow it absolves the actions, inaction, mistakes, or oversight that occurred during the incident. We don't always have control over our staffing and budgets, but let's be honest; most times we have control over how we operate on scene.

We (firefighters) don't like pointing fingers or placing blame. For so many years we accepted that we will lose our brothers and sisters, and accepted that we will tragically change the lives of their families from the day of the incident forward. You have no idea how our attitude, our acceptance of death or serious injury as a risk of doing business, negatively affects the families left behind. Our public display of grief ends, our flag draped despair fades, and the families are still left without their loved one . . . forever. As I look back at the over 25 years of being a firefighter for the same department, forever is still forever. I will leave here, I will move on, and they are still gone . . . forever.

We should never accept death as a risk of doing business, unless that risk was truly in the effort to save another life. We are all familiar with the buzzwords "risk a lot to save a lot, risk a little to save a little." You need to really live with this mantra every day, and I mean every day.

I started with this introduction because I am stalling; it still isn't easy to write about or talk about the Pittsburgh events that have changed this firefighter, line officer, chief officer, forever. Seven years into my career, the Bricelyn Street fire occurred on February 14, 1995, in which Pittsburgh lost three firefighters—Tom Brooks, Patty Conroy, and Marc Kolenda. Ebenezer Church nine years later on March 13, 2004, was where Pittsburgh lost two firefighters—Charlie Brace and Rick Stefanikus. I was not on scene at the Bricelyn Street fire; however, I lost one of my best friends to that incident. I listened to the incident as it unfolded, and I knew that I had lost my friend. I was a pallbearer who helped carry Patty's casket to her grave site.

I was at the Ebenezer Church fire when the collapse of the church steeple killed a firefighter and a battalion chief. This chief was one of my first captains, a chief of mine, and he was my friend. I had to go to his home and tell his wife that he, most likely, would not be coming home, that he was buried alive and we couldn't find him. She most definitely already knew this when I pulled in front of her home in a chief car that was not her husband's. I knocked on the door; she peeked out of a small draw of the drape and closed it very quickly. She didn't open the door; it was as if by not opening it, this would change the tragic turn of events. It wouldn't; it didn't. I was a pallbearer who carried Charlie's casket to his gravesite.

What was painstakingly hammered home from the Bricelyn Street fire, and from being an escort for the National Fallen Firefighters Memorial service, is that as a company line officer I wasn't only leading my crew into a structure fire or dangerous situation; I was taking their spouses, their children, their parents with me. Some fire service personnel think that is too heavy a burden to give to an officer, that you can't concentrate on the task at hand by carrying that additional burden to your emergency decision-making thought process. I disagree.

Pittsburgh had historically been a very aggressive, union, interior attack department. For *criminy* sakes, they are IAFF Local 1 and very proud of it. That being said, I cannot imagine that if I didn't think things through quickly, how was I going to look into the eyes of a widow or widower if I had not made decisions based on the information and sound judgment for the situation? I will not say I was perfect. Several times during overhaul, in the light of day or with artificial light on scene, I found myself asking, "What the hell was

I thinking taking my crew here?" I can say that trying to make the proper decisions, with all of the information in mind, I made much better decisions for the situation, my crew, and the incident.

Between me and you, I know this . . . for those left behind . . . the family of the fallen, they want to know, they want to talk to the people who spent those last minutes with their loved one. They want to know what were the last words, the last thoughts, and the state of mind. *I led my crew so that I could lovingly and confidently let my families know that I tried to do the right thing.* I know that as an officer I could in all good conscience tell them that we looked out for them; we did everything we should have, and we loved them too. You can't do that if it's a lie. You can't talk to or look into the eyes of the wife or husband or the children if it's a lie. You can't do that if you know that you didn't follow the right rules, standards, or guidelines.

I learned a lot about incident command through all of my classes at the National Fire Academy (NFA) and elsewhere. I learned a lot from my experience as a company officer. If you become an officer, you are to lead, not blindly follow. Being an officer is more than a badge, a bugle added, or a pay increase.

I was not the incident commander (IC) for Ebenezer Church, but as a new battalion chief, I learned that as an IC of an incident or major structure fire, ICs don't lose sight of the situation until the end of their command. They don't let others interfere unless they are willing to formally take command of those who I have known and loved.

So many fire families pray that we, as a service, let their loved ones' loss be a lesson learned for us. It is no cliché to say they don't want their loved ones to have died in vain.

Hello fire young firefighters, officers of the fire service . . . don't let their loved ones, our loved ones, die in vain.

Don't forget:

1. As an officer, particularly a chief officer, trust your own instincts. Firefighters want to fight fire many times to their own detriment; don't listen to the "We got this, Chief . . . just a few more minutes" when your standard operating guidelines, experience, and gut tell you otherwise. Pull them out.
2. Let your crews know that you want them to report the conditions and hazards they see, that you want as much information as possible to make informed decisions.

3. Safety and accountability are every single firefighter's responsibility for the success of the mission.
4. Command is held until everyone has been accounted for and has left the incident.
5. Command has only one recognized voice, even if a unified command is used.
6. Take time to choose who you want to emulate; the officer wearing the dirtiest clothes, with the most charred helmet, may not be the one.
7. Don't be so blinded by your rank that you don't see when a firefighter may have the solution to your problem. Make sure they know it if you find that they do.
8. An engineer/driver can make a line officer look very, very good, or very, very bad.
9. Beware of the loaded question. Don't be afraid to say, "I will get back to you."
10. It's not cliché; never forget the main goal—everyone goes home.

Bill Webb

Crawling Down the Halls? Try the Halls of Congress for a Uniquely Different Challenge

A former member of the first Bush administration, Bill Webb served as a special assistant to cabinet secretaries at the U.S. Departments of Education and Labor, preparing briefings and accompanying the secretaries as a personal travel aide. He is also the executive director of the Congressional Fire Services Institute (CFSI). In representing you and coordinating with literally every national fire service organization, Bill "crawls down the halls" working closely with members of Congress and fire service leaders on developing federal legislation and enhancing federal programs designed to help us. Bill also serves as the vice chair of the National Fallen Firefighters Foundation board of directors. While not an active firefighter as such, there are few as sharp and as politically astute as Bill is in truly understanding our needs. There are several contributors to this book who aren't firefighters, but who support us and help us—sometimes doing it better than we can ourselves. Bill Webb and his crew do that every day—and he is an expert in passing it on, to benefit us.

Vice President Joe Biden and Bill Webb share a moment. VP Biden is well known for his longtime, proven support of the nation's fire service.

Politics is Local: You Can Make a Difference for Your Fire Department

Firefighters have a tremendous number of responsibilities—responsibilities to their departments, to their families, and to themselves. To a large extent, their safety and the safety of their brothers and sisters is determined by the decisions and actions taken in the station and on the fireground. But there are other places where firefighters—whether career or volunteer, chiefs or rookies—can have profound impacts on making the fire service a safer profession for all. One such area is in legislative halls both at the state and federal levels. Because I have spent the past 18 years working fire-service-related issues in Washington, D.C., I'd like to focus my thoughts on advocating issues on Capitol Hill and why every firefighter should get involved in this effort.

In 1999, I attended a meeting with a group of fire service leaders and a staff member for Congressman Bill Pascrell in the congressman's congressional district in northern New Jersey. The purpose of the meeting was to develop a proposal for a major fire service grant program. At the time, the federal government offered only a minimum amount of support in the form of grant funding to the fire service. There were some funds available through the Community Development Block Program, approximately $2 million through the Volunteer Fire Assistance Program, and a very small amount of other funds within various federal agency budgets.

The grant program proposed by this group was endorsed by Congressman Pascrell, who in turn introduced legislation that became the precursor to the Assistance to Firefighters Grant (AFG) program. Enacting federal legislation is not an easy process regardless of the merits of a proposed measure. This was certainly the case with the Fire Improvement and Response Enhance Act (FIRE Act). It took the collective efforts of all the major fire service organizations, a bipartisan group of congressional leaders and hundreds—if not thousands—of local fire officials to get the bill approved.

Former Speaker of the House Thomas "Tip" O'Neill once said, "Politics is local"—three simple words—yet they are considered to be the golden rule of understanding what makes Congress act. There would be no FIRE Act or *Staffing for Adequate Fire and Emergency Response* (SAFER) program if local fire officials didn't contact their members of Congress and urge them to support these two programs. On a strategic level, members of Congress who served on the committees of jurisdiction when the FIRE Act was first introduced were being inundated with calls from their local fire officials. The national organizations shared talking points with local fire officials throughout the process to make sure consistent messages were being delivered to our lawmakers. Mounting pressure from fire officials throughout the country finally forced Congress to act on the measure. And here we are, many years later with both AFG and SAFER having awarded billions in federal grants to local fire departments—and the programs are still in place.

Prior to passage of the FIRE Act in 2000, the common complaint from the fire service was that law enforcement was receiving all the money. Back then, that was true. But that's certainly no longer the case. The fire service—thanks in large part to the ground troops who have reached out to their members of Congress and engaged them in our issues—is a political force on Capitol Hill. Working together, the fire service collectively (career, volunteer, chiefs, instructors, investigators, industry, and so on) has prevailed on a number of issues. Though the dollar figures don't always tell the story, our efforts have made a difference on a number of issues, including not only the sustainment of the grant programs, but also the reauthorization of the United States Fire Administration, additional spectrum set aside specifically for public safety, and improvements in the Public Safety Officers Benefit Program.

Yet we should never become complacent or rest on our laurels. As long as sequestration and future budget challenges hang like a dark cloud over Capitol Hill, local fire officials must remain vigilant and engaged with their members of Congress. Remember that our issues are not partisan, they are nonpartisan—public safety issues that have a direct impact on your member's constituents. You should feel a sense of obligation to your department, the members of your community, your families, and yourselves to educate your elected officials, explaining to them from a local perspective the benefits of federal programs.

Yes, politics is local, and nobody can explain the local perspective better than you. So exercise your right to speak, and encourage your peers to do the same. That's how the fire service can continue to prevail on Capitol Hill.

Mike Wilbur

Mike Wilbur has been a volunteer firefighter for almost 40 years and a member of FDNY for more than 30. He recently retired as lieutenant of 27 Truck in the Bronx, was previously in Ladder 56 for 15 years, and before that served 8 years as an apparatus operator. He was on the FDNY apparatus purchasing committee and personally crafted, proposed, and led the recent FDNY seat belt retrofit project—a project that has already led to the saving of lives. Mike is a contributing author for most fire service publications, served on the IFSTA validation committees for the apparatus manuals and the USFA's Emergency Vehicle Safety Initiative. He specializes in emergency vehicle operations, apparatus placement, and apparatus purchasing, which translated means: Mike loves fire apparatus! I consider Mike one of my best friends and one of the most passionate firefighters I know. A man with a heartfelt vision, he has secured support and grant funding to completely change apparatus design—specifically related to seating and securing each of us. He has lived a life of generously passing it on, and here he is again.

Mike Wilbur

Six Lessons for Success

As many of you read this, you are early in your careers in the fire service or perhaps just starting one. For me, after more than 31 years in the New York City fire department serving in the borough of the Bronx in a couple of busy ladder companies and 39 years as a volunteer in upstate New York where I live, my time in the fire service is winding down. However, over that combined 70 years of experience I have learned a few things that I would like to pass along.

I received a burn only once, after my third fire in 24 hours. My gloves were wet, and I received a steam burn through my glove while shutting off an acetylene cylinder at a working fire. I never let that happened again. I always had a dry pair of gloves; lesson learned and I got away easy with a sunburn.

You might ask why I never received serious burns like some of my colleagues. I might have just been lucky, but I would like to think that over time I learned how to do a quick, thorough size-up and always had good situational awareness. And I always took the job seriously. I cannot tell you that I was never complacent, but I really strove not to be, especially as I gained more experience. After we received an alarm, I always put my PPE on the same way—*the right way, all of the time!* Whether it was a first-due working fire in December or a water leak in August when it was 100 degrees, I wore my PPE the same way *the right way, all of the time.* I did not want to get caught in a flashover and burn to death. I did not want to be *that guy* who ends up in "Firefighter Close Calls" or as an LODD statistic. Lesson 1: wear your PPE the right way, the same way, all the time.

Having done extensive work in emergency vehicle driving and seeing firsthand the aftermath of a firefighter ejection line-of-duty death, I always wore my seatbelt whenever I could both as a driver and an officer. Lesson 2: wear your seatbelt. Why? During my career, I saw the difference between firefighters who took the job and the responsibility for their personal safety seriously and those who did not. As a company officer my worst day was when I had to go to my friend's front door and tell his wife and kids that he was not coming home. I hope you *never* have to do that. He was a firefighter who did take his own personal safety seriously; however, he died at the World Trade Center where, in that case, it did not matter much. But if you choose not to wear your seatbelt, if you choose not to stop at red lights and stop signs, if you choose to eat another dozen donuts, if you choose not to take care of yourself or your health, at some point in your career you could end up killing yourself. Why? Because you did not take your own personal safety seriously. You will be dead, the future will not matter to you, but you will leave behind a community and a fire department, and your family will be in a state of ruin. Psychological ruin, perhaps financial ruin, and shock, because you did not take your responsibility for your personal safety seriously. You responded to be part of the solution, not become part of the problem. Lesson 3: *you* are responsible for your own personal safety. Take it seriously; your fire department trains you for it, your community expects it, and your family deserves it.

Speed kills! Whether you are driving to the fireground or you are operating on it, speed kills. I drove an aerial ladder for 10 years in the Bronx and drove to numerous volunteer runs. I wish I could tell you that if I drove 80 mph instead of 50 mph that I would have saved five more lives; however, I can't tell you that because it is simply not true. If civilians were alive when we showed up, we brought them out alive; if they were dead, half of them died before the fire department was ever called. As a company officer, I watched a firefighter put on his breathing apparatus prior to our arrival to a working fire. I then watched as he ran into the building as if he had a rocket on his back. When we reached the door to the fire apartment, I asked him what floor we were on. He said he did not know. Wrong answer. Why did he not know where he was? Because he never did a proper size-up and just ran blindly into the building. It has often been said that firefighting is the war that never ends. How can you fight a war when you do not know where the enemy is, how big it is, or where it is going? Size-up and situational awareness will give you the answers to those questions. Lesson 4: *slow down!* This will allow you to do a complete size-up and maintain situational awareness.

Be professional! Professionalism has nothing to do with collecting a paycheck. Professionalism has everything to do with how an organization is run and how its members train and prepare for fires and emergencies. I have been to some career places that lacked professionalism and have been to some volunteer fire stations that fit the very definition of professionalism. Remember—running around on the fireground yelling, screaming, acting like a chicken with its head cut off, and behaving worse than the civilian whose house is on fire has never put out a structural fire. What puts out structural fires are training, experience, and professionalism. Be professional every time, everywhere that you wear something that says FD on it. You are representing the organization that appears on that shirt, and everyone has a camera. Lesson 5: be professional; act professional.

I was fortunate in my career to be promoted. Passing a promotional test or winning an election simply does not an officer make. My take on becoming an officer may be a bit different from what you might learn at an officer's school, but this has been my experience. I believe that with my experience and reputation as a firefighter, plus my evaluation by my company in the first few months, my crew basically put their trust in me, their lives in my hands, and allowed me to lead them. They listened to me. One of the greatest compliments that I received while working in the FDNY at the Cross Bronx Express is that the people wanted to work with me because they knew I would not get them hurt and that they would go home at the end of the tour. Because I had the privilege of working with some of the best interior structural firefighters and officers in the world, that comment meant the world to me. If you enforce every rule, and you constantly draw the line in the sand, at the end of the day as a company officer you probably will become a very unpopular and ineffective leader. But being a company officer is not about being popular. I believe that you may not always be popular, but you will be respected for making the tough decisions, which will lead to being the most effective leader. Lesson 6: lead by example; do what I do, not what I say; and treat firefighters with dignity and respect. You will get that back tenfold. Remember you are working with and leading the greatest bunch of people in the world, this nation's firefighters.

JANET WILMOTH

Making Us Feel Just a Little Uncomfortable

I've been fortunate to become friends with most of the major fire magazine editors in the last 40 years. Janet is the former associate publisher of *Fire Chief* magazine (*Fire Chief* magazine went out of business as of this writing), and if you followed her columns or blogs, you know she is never short of an opinion. Sometimes "nonfirefighter" opinions rub us the wrong way. Sometimes they should. Growing up in a firefighting family, Janet has been a friend to the service since she got started. She has written about issues that we would rather ignore. Fifteen years ago, Janet created "In Service" for *Fire Chief* magazine to address issues with apparatus safety and maintenance. For years, Janet passed it on each month as she wrote her thoughts in both the magazine and the web. Disagree with what she writes? Hell yeah, I have on several occasions. But Janet was always good at making us think hard, something important for all of us. I look forward to her continuing her "thinking" role in the near future.

These are my brothers—Battalion Chief Dan Cook (left), me, and Bureau Chief Don Cook. They are twins, both retired from Lisle-Woodridge, Illinois, fire district and have been great resources and inspiration for my work.

Five by Five

Can you give me five minutes? I've got five things I'd like you to think about, whether probie, firefighter, or seasoned fire chief.

For more than 25 years, I have had the unique opportunity to observe the fire service and voice opinions, sometimes from a global perspective, but mostly from North America.

I believe that the past 25 years have been some of the most exciting times in the history of the American fire service. Not only have standards been established for the professionalism of the fire and emergency services for both volunteer and career departments, but the adoption of new technology has blown away the old adage describing the fire service as "200 years of tradition unimpeded by progress." Rubbish! There will always be laggards in every industry and organization, but today's generation of emergency responders is quick to use new and improved tools and techniques to their advantage.

From my perch, there are five areas that are worth asking you to think about:

Restrain yourself. Perhaps if we call it by another name, firefighters will be more inclined to wear seatbelts. It's truly sad that every year firefighters die needlessly because they simply didn't buckle their seatbelt. The real heartache is for the families who survive these fallen firefighters, knowing they did not need to die. There really is *no* excuse not to buckle up. Please don't let anyone in your vehicle, shift, or department ride unbuckled. Real heroes wear seatbelts so they can arrive alive and return to fight another day.

Clean it up. Inside your gear and out. Every firefighter should take a shower before going home. Leave work at work, and that includes exposure to carcinogens, hazardous materials, even bedbugs. Research will soon support the importance of personal hygiene, just like current research supports regular cleaning of your PPE—helmets and hoods to bunker coats and boots—to reduce the risk of cancer. Don't "wear your work" into the day room or to your living room at home.

Download this. In the emergency services, you really are on duty 24/7, even if you're a volunteer. What you do after hours will reflect on you as a public servant. Technology has opened the doors of information to our private lives that we can never close again. Social media and smartphones thrive on news. Someone is always watching, ready to text or video. Oh, and be cautious what you post.

Mayday. It is important that you call "Mayday" for help on the fireground, but it's also important to ask for help when you need help in your personal life. Firefighting and emergency response are stressful professions, and years of stuffing it inside is just not healthy. Don't wait until your family or life is falling apart to get help from a trained healthcare professional for emotional issues. You go to the dentist with an abscessed tooth or to a doctor with a broken arm; emotional pain needs treatment, too.

Common sense. One of the things I find most exciting about the future of the fire service is that research and new technology are proving that *common sense* on the fireground has not gone out of style, and it does work. Traditions are important, but some need to be filed under "history."

Fighting fire and responding to calls for help is a noble profession that serves mankind. Never stop learning, and never stop taking care of yourself, because ultimately you are responsible for yourself.

I am honored to serve this profession.

Arlene Zang

The Loss of a Firefighting Daughter

Robin (Zang) Broxterman was a 17-year captain and veteran with Colerain Township FD (Local 3915) near Cincinnati, Ohio. Robin and part-time firefighter Brian Schira were part of an engine company that responded to an early morning alarm. Arriving at the scene, Broxterman and Schira went inside. "Making entry into the basement," Broxterman said via radio. "Heavy smoke." Minutes later, firefighters told a commander to order Broxterman and Schira out of the house, because "conditions are changing." No response. The rapid intervention team members battled their way into the house. They found Broxterman and Schira in the basement. Their deaths rocked Colerain's 180 firefighters like never before. The parents and family of Brian Schira became parents of a fallen firefighter—nothing they ever expected. It also rocked Robin's fiancé, Don Patterson—a brother firefighter as well—along with Robin's daughters, Sierra and Courtney, and especially her dad, Don, and mom, Arlene Zang. Robin's mother, City of Cheviot Fire training officer and retired Green Township firefighter/paramedic (20 years) was eager to share her thoughts.

Robin's daughter, Courtney Broxterman, receives her mother's flag.

Brother, Where Art Thou?

Fire departments are very good at planning funerals. They honor the fallen firefighter with great tradition and utilize every resource to ensure a service worthy of their hero. But how do they care for the remaining family members who struggle to celebrate Christmas, birthdays, and other memorable events without their loved ones?

I've watched those brave survivors struggle with the finality of death and the anniversaries that follow. I learned how they felt, the hours of crying, the pain, and how they got through those really rough times. But some of that pain was unnecessary. Thoughtless words and actions by a few are forever etched in the memories of those survivors. Unfortunately, those few were fellow fire department members, the very same people who vowed to support the survivors. I was shocked and appalled as I listened to survivors relate the many ways the fire service failed them. In some instances it was merely a communication breakdown and there was a plausible explanation, but in other cases their actions were difficult to defend.

One family received their annual Christmas party invitation. But upon arrival, they were told the party was limited to current members' families only. The wife of the fallen firefighter was left standing at the door, trying to explain to her two small children why they could not attend the party that their family enjoyed for many years. A 12-year member of the ladies auxiliary was told to resign because membership was restricted to firefighters' wives only and that she no longer qualified. Another widow heard on the radio about a memorial service honoring her husband. On arrival was told that this service was for firefighters, not family. They did allow her to stay but placed her in a rear pew.

The chief called one widow and requested that she return her husband's badge and uniform. She offered to purchase it, but was told selling "department property" was against policy because she might impersonate a firefighter. One township dedicated their new firehouse to their fallen hero and invited his family to the parade and celebration. Two years later, they re-dedicated the same firehouse to a war veteran. The family read about it in the newspaper.

"Remembering good times together." Posted on the department web site, just days after the firefighter was killed in a fire truck accident, were pictures of the firefighter sleeping in a chair, beer cans on the table. Just imagine the comments that family had to endure. No alcohol was involved; they other driver ran a red light. But why read a 60-page LODD report when one picture tells the whole story?

Other actions were downright cruel. One mother was told her son's death was his own fault because he chose to work the overtime shift. At another department, members were ordered by the chief not to attend the memorial, no explanation given. Still another surviving family was told that the accident pictures were too gruesome to view, but the firefighter's children got to see the pictures in living color on the six o'clock news. Someone within the department gave pictures to the news, but said, "I didn't know they would be on TV!"

Many departments have refused to convey the circumstances surrounding their firefighter's death. One department told the family to get an attorney if they wanted answers. Another chief hosted a seminar on their firefighter's death, invited area firefighters/departments, but refused entry to the family. Several departments have refused to process paperwork to file a Public Safety Officer's Benefit (PSOB) claim. If you don't know what that is or how to file, there is a local assistance state team available to help file for federal, state, and local benefits after a line-of-duty death. Information can be found at www.firehero.org.

Robin's daughter, Sierra Broxterman, rubs a brick with her mother's name.

One family praised their son's department, told how the firefighters stood by them for five days, chauffeured them, and promised to never forget. When a member of the department spoke the words, "Brother, rest in peace, we will care for your family now," the family felt so loved. But after the funeral, weeks, months passed with no contact. No Christmas card, not even a phone call to ask about the children. What makes these revelations even more outrageous is that most of these survivors were involved with the fire service for many years. This truly was their second family. Those firefighters gave all, and their families deserve better.

Firefighters are used to making critical decisions every day. They do a great job of improvising, and their motto is "adapt and overcome." So why do firefighters, chiefs, and entire departments become brain dead after the LODD services are over? I would like to believe this is just because they are not thinking. Perhaps those in the departments are also still in denial and they just don't want to be reminded. Or if the situation makes them uncomfortable, they tend to fall back on the rules for excuses. Chief, did you really think that widow was going to put on her husband's size 52XXL uniform, flash his badge, and pull over cars at midnight? Truth be told, most survivor don't even have the will to get out of bed in the morning. This is so wrong! Even if guided by policy, it is wrong. For surviving families, there are no rules. What seems logical to you may be perceived by the family differently because survivors are extremely sensitive. Perhaps the perception of the deceased family may be a bit off, but to these surviving family members, perception is everything.

I am confident that most times the fire department tries to do the right thing for families. Even when things go wrong, I am not convinced that it is entirely the department's fault. Chefs use recipes to ensure a tasty outcome; athletes have rulebooks; maps are consulted before vacation . . . but there is no guidebook for supporting the survivors after a line-of-duty death; not one magazine article or even one page written to focus on some of the critical issues faced by those survivors, months or years after their loved one's death. I don't believe a one-size-fits -all book could be written because each situation is different, but I can offer two simple words: *be there!* That family heard you in church when you spoke about remembering. Those survivors really believed you when you uttered all those words about being family.

Educate your fire department. Invite a survivor from the National Fallen Firefighters Foundation to speak. If you find yourself in a tough position, call a survivor; they may be able to offer you some options or in some circumstances may even be able to run interference for you. Consult your chaplain, or have a psychologist explain the grief process. If all else fails, *use common sense*. Put yourself in their place. How would you want your family treated? All the survivors want is for the department to honor their loved ones and acknowledge that their family, especially their children, still exist—like you promised in that glowing eulogy.

Department names have been eliminated to protect the guilty, but if you see yourself or your department above, it's not too late. Extend an invitation to the family, make a phone call, or send a card. Take the kids to a picnic or ball game or attend their graduation. Invite them to a fire department function. Remember that the firefighter who died did not resign or retire from your department. But most important, *remember that you called them family*. Treat them like it.

The American flag, presented to every LODD family, at the National Fallen Firefighters memorial service, is the fire service's promise never to forget.

FF/Paramedic Brian Goldfeder and FF/EMT Dave Stacy . . . Some Perspective from Some Improved Chips off Some Relatively Old Blocks . . .

While I sorta wanted my son to be a firefighter (I knew my daughters Amy and Dani did *not* want anything to do with it), it wasn't an obsession of mine. And that's a good thing because that was not on his agenda. So off he went to Ohio State University. About a year or so into it, he came home one day and we had lunch together. "Dad," he said, "I looked at the catalog of majors at Ohio State and I really don't see much I'm interested in."

Not being able to find a major at OSU is like not being able to find a sandwich in a NY Deli. So we talked and I asked, "So what do you think you do want to do or be?" . . . and he answered . . . "Well, I thought about being a firefighter."

WTF!? *(Insert fireworks, a 200 piece band, skyrockets, etc., here please.)*

I had no clue that was coming but was certainly thrilled to hear it. So after some research, meetings, interviews, and related stuff, he was accepted to the University of Maryland—and was also accepted as a probie at one of Prince George's County's (Maryland) volunteer companies, the College Park VFD, Co. 12.

While I could probably write chapters about the College Park VFD Company 12 and the many other volunteer FDs within the PGFD, I want to save that for another time. What I do want to share is the fact that for nearly four years, Brian was a "live-in" firefighter/EMT and gained some amazing life and fire service experiences. He lived in the CPVFD "sackroom."

Their station is located on the campus of the University of Maryland, College Park. Located directly above the apparatus bays, the sackroom is designed like a campus dormitory. Each semester, 18 full-time student volunteer firefighters live in the sackroom, the John L. Bryan dormitory. In exchange for their housing, they work three to four duty shifts per week. While they can ride as many calls as they like, their duty shift is required riding time. As the core of their operations, each student volunteer makes several hundred responses per semester. They serve in all capacities, including firefighter, EMT, ambulance driver, engine driver, truck driver, line officer, and staff officer.

It was in the sackroom that Brian met his roommate, Dave Stacy, who also comes from a firefighting family, including his father James, who served for more than 32 years with the City of New Haven, CT, Fire Department. Dave and Brian became good friends, made thousands of runs together, and remain close today. In addition to school and the firehouse, Brian worked at the Maryland Fire and Rescue Institute, the state fire academy, while Dave had several internships, including participating in fire research at the National Institute of Standards and Technology (NIST). Within a few years, both graduated with their bachelor's degrees from the University of Maryland. Brian tested and was hired by the Prince George's County Fire/EMS Department (as of this writing he is a firefighter/paramedic). Dave tested and was hired in the Durham, NC, Fire Department, where he serves as a Firefighter/EMT. Both married with kids, their careers, friendship, and families continue to grow. Both are doing so with lifelong appreciation for what they experienced and learned at Company 12.

Brian and Dave operating in the back of ambulance 129. The CP VFD ran a wide range of EMS incidents.

I asked them what they would like to pass on to probies and new firefighters specifically. Take a look:

- Respect is earned by actions, not words. Keep your mouth closed and ears and eyes open. This will hold true throughout your career.
- Small problems should go to senior members, not the company officer.
- There are no dumb questions. If you do not know something, figure it out. If you can't find an answer yourself, ask someone. Your crew would rather you ask a question, even if you should have already known it, than not ask, and be clueless on an emergency.
- Your reputation starts the first time you meet your crew and members of your department, and you should try to meet your crew the shift before you start a new assignment.
- Your appearance plays a big part of how people perceive you, right or wrong. Think, if your mother, daughter, or father were sick or injured, would you want someone like yourself there to help?
- Stay off your cell phone throughout the day unless there is an emergency, for at least the first few months on the job.
- Take pride in all that you do on the job; your crew will notice the little things.
- Don't wait for someone to tell you to do something. If it needs to be washed, wash it. If that trash needs to be taken out, take it.
- Don't be the last person on the truck. And yes, they will leave you if you're not ready.
- Always offer to take over a chore from a senior member. Even if the answer is no, the effort will be noticed.
- Make sure the coffee is prepared at night before you go to bed so the crew can make a fresh pot in the morning with little effort; make sure the TV is off and dayroom/sitting area looks neat before you go bed.
- Even if you don't drink coffee, make it! You'd be surprised how far a fresh pot of coffee goes for the crew in the morning.
- Your first day on a new shift, especially on your first station assignment, bring breakfast. (Breakfast from a real bakery will go much further than Entenmann's, trust us.)
- Know what you keep in your pockets, where, and why. You don't want to fumble through six pockets to find a screwdriver on a bells call, and

you especially don't want to fumble through them when it hits the fan and you need something now.

- Love the job and work as much as you can, but don't forget you've got a family at home that wants to see you too.
- Love it or hate it, EMS is here to stay and (probably) part of the job description when you joined/got hired. No one is asking you to become a doctor, but at least try to be a competent, compassionate EMT or paramedic.

Brian and Dave operating on the roof of a garden style apartment during a multi alarm apartment fire.

Both Brian and Dave believe that, aside from responding and training together, the firehouse brotherhood includes many meals together.

Anonymous

Several years ago, a veteran fire chief wrote a love letter of sorts to his son, who was entering into probie school. Here are that chief's personal thoughts.

Advice for the Brand New Firefighter . . . Well, Not Really a Firefighter. Not Yet.

Dear Son,

First of all, you are not a firefighter yet. You're a fire trainee, a probie, a rookie, or whatever your area calls it. You are *not* a firefighter.

In addition to reminding you that you have very few rights yet, you should know the fire service is *not* always a democracy, we are *not* interested in your opinion, and you need to shut up, listen, work, and *train*. So, *what else* can I tell you? Perhaps the fact that you now must *do exactly what you are told*, and you may actually be just a little offended or upset by some of the things you hear or see. If you aren't able to handle that, this may not be the job/career or volunteer opportunity for you. Maybe the part about you having to do more work in the firehouse than you do in your home will send you into a tizzy. If not, read on.

And by the way, for you non-probies out there, don't get offended; when I say the new kid (boy, girl, whatever gender—I couldn't care less) has few rights, I'm talking about the fact that when they are *new*, they need to listen, learn, and get ready for the best job ever (career or voluntary), which means they have the right to shut up and listen for a change. They have the right to learn what teamwork is. They have the right to get used to be told what to do. And they have the right to do their own laundry, clean their own apparatus, and prepare meals—without paying someone to do it. And they have the right to get as much training as possible so they increase their chances of "going home." But they can't do that if they don't do the above—or the below.

So, newbie, as you get ready to start, here are some thoughts, ideas, advice, and related little gems of knowledge for you . . . in no particular order. Some of these are originals, and some I picked up from other fire veterans, but it doesn't matter—this information might just help you.

1. When getting ready for duty, check your equipment—and perhaps other equipment, if that's your department's tradition or rule—but *always* check yours. Make sure you *always* check your SCBA; if it doesn't work perfectly, you can't breathe, and that sucks. When the tones go off, it's too late to make sure you—and your equipment—are ready. Do it now.

2. When reporting for duty, ask if anything needs to be done, or look around and *find* what needs to be done. The newer you are in the department, the more you need to be in the truck room looking, touching, and learning.

3. Make sure all your personal gear is ready to go. Every time you come on duty, check your pockets for your gloves, light, rope, and so on. What should you carry in your pockets? Some things I carry include a rope, a flashlight, a flathead screwdriver, a Phillips-head screwdriver, and a wire cutter. Also consider carrying a Leatherman tool, webbing, modified channel lock pliers, and door chocks. Why carry all that? You'll understand why very soon.

4. Make coffee . . . if that's what they drink at your firehouse. Me? I can't stand the stuff. Makes me hyper, and you know how I get then.

5. As for meals and cleaning, find out what has to be done. Peel the potatoes, cut the onions, or even cook the meal. Fill or empty the dishwasher anytime it needs to be done, and clean *whatever* needs to be cleaned. And by the way, clean the toilets (no kidding). Here's a tip: Don't flush the toilet once it's clean; leave the clean, soapy water in the bowl. This shows the other firefighters the toilet has been cleaned. *How about that?*

6. Volunteer to do whatever needs to be done. *Whatever.*

7. After eating, be the first to the sink to scrub the pots, pans, and dishes, and get it done very well—as if someone is going to inspect how clean the stuff is. Someone will!

8. Be proud that you're going to learn to help people who are having the worst day of their lives. Few people get to do that. And very few people are firefighters—*real* firefighters.

9. Learn your firehouse or department policies and traditions—and know them by heart.
10. Always say "Sir" (or "Ma'am," if your "Sir" is a woman) to chiefs, officers, and instructors. You could also say "Chief," "Cap," "Lou," etc. They've earned and deserve your verbal and action-related respect.
11. When arriving at a fire, always get off the rig with your SCBA and apparatus tools. Some people won't always get off with their equipment. Don't be one of them; they are lazy, and lazy is dangerous. You can always take the equipment off or set a tool down, but it takes longer to put it on or to go back and get something if it's urgently needed. Be ready. Things can turn to s#%* quickly. If you and your crew are ready, it can matter. If you aren't, it can suck.
12. When on a run, listen for your officers' instructions and *follow them*. Always pay attention.
13. Always stay with your company. If you freelance or wander, it can kill you or other firefighters.
14. At fires, *stay low*. Heat and smoke rise. Listen to the sounds of what's burning, hoselines operating, other firefighters and, most importantly, your partner. Plus, know how to feel for the heat. Study and learn fire behavior, and get as much hands-on training as possible.
15. Keep your mouth shut and your eyes and ears open . . . *always!*
16. *Never* give up. Everyone makes mistakes. Everyone screws up. Me. You. Everyone. Don't be afraid of the mistakes and screw-ups; they'll happen no matter what you do. What you can do: everything in your power to limit the seriousness of the screw-ups. Ask questions when you don't understand. Ask questions if you *think* you understand. Maybe even ask questions when you get it, because chances are the probie next to you doesn't get it and won't ask.
17. *Listen!* (I know this and some other items here are repeated—maybe that's for a reason.)
18. *Learn, study, drill, and train, train, train.* It never ends. Also, it doesn't hurt to be assertive; however, it can hurt you to be *too* assertive. Jump on things like I already mentioned, such as dishes, bathroom details, and other firehouse duties; but, don't be the rookie who's kissing butt just to get a good review or to fit in. Besides, you'll stand out to the other firefighters and get labeled. *Let your skills and actions speak for you; you will be judged by your actions.* It's good to try to be No. 1.

No firefighter would want to be in a fire with anybody who's comfortable being less than the best they can be. *Learn, study, drill, and train, train, train.* It never ends.

19. Respect those who have done this job before you. That's nearly *everyone*. And don't get comfortable. I've seen a lot of probie firefighters in the final months of their probation get waaaaay too comfortable and forget their place in the firehouse, especially when there are newer probies under them.

    Don't be a 6-22; that's someone who has 6 months on the job but acts cocky—like they have 22 years of experience. Stay active, stay in the books, study the tools and equipment, and work harder than the hardest-working firefighter—but *not* in a showy, notice-me way. Your last few months of probation will last forever if you screw up. No matter what the problem is, don't forget that this is the best job around, and you don't want to lose it. There's no such thing as a ranking probie. Be respectful, stay humble, shut up, train, and learn.

20. Take care of yourself. You are priority no. 1. Be *safe*. *Constantly* think about safety, and remember that the job of the fire department is to help people with a problem while *not* becoming part of that problem. This is a risky job. Sometimes we *must* take risks, but do your best to study, train, and understand when the risk is worth it, not worth it, or just dumb.

21. Feel like you earned your seat on the apparatus. *Earn* the seat. *Earn it!*

22. Know that the public, especially kids, watch you and look up to you. What do they see and hear? Act as if your fire chief is sitting on one of your shoulders and your mother is on the other.

23. Ask your boss about your progress in private. What are you doing right and what are you doing wrong? But don't ask in a suck-up way. Be professional.

24. What you do when you first start out will set your reputation and follow you throughout your career. If you don't start out on the right foot, the department will show you the door. The crew knows more about you than you think before you even show up. And if you have a MySpace, Facebook, or one of those expose-yourself accounts, assume *every* member of your firehouse has already seen it. Bet that made you happy, Pookie. And, one more thing: Stay off those stupid Internet fire-related chat rooms, bulletin boards, and rants. Most of those sites will expose you to nameless clowns who prove that many

in this business have forgotten (*or never knew*) what brotherhood/sisterhood is.

25. You're a probie. Don't get too stressed over that. Just keep your mouth shut, work hard, train, study, train, study, be cordial, friendly, and humble. You have no opinion until you earn it. And you can't force it. That will come with time after training and gaining experience from runs.

26. Leave your cell phone in your car until a time (months from now) when all your duties are complete. I repeat: Stay off the phone, off the instant messagers, off the text messagers, and focus on your job. You are going to be responsible for your life and the lives of other firefighters and civilians. Focus on that.

27. Before you arrive for your first day, stop at a nice bakery and pick up some dessert to take to the firehouse. Get a cheesecake or something really tasty. Or maybe bring in something homemade if you know someone who can bake.

28. When you're expected to be at the firehouse, fire training, the academy or wherever, *always* arrive very *early* and ask—or just do—what needs to be done.

29. There will be certain duties assigned to each day of the week. Tuesday could be for cleaning the trucks, Saturday for cleaning the building, and so on. Keep track and know the plan. Stay busy around the firehouse.

    Look in *all* apparatus compartments, and *memorize* what's in each one. When on duty, always be ready to get on the rig and respond in the seat and position to which you're assigned. *Memorize* your apparatus duties and what's in each compartment. When you're taught about a particular tool, become an expert in that tool. If you aren't an expert, who will be?

30. The senior firefighters at your station might have *their* place to sit at the table or in front of the TV, but don't do that. Besides, you probably have probie tests to study for. I repeat: Don't sit in front of the TV. No more video games, either. No matter what the atmosphere, you're being watched, so study! Or don't, and just go work at a bakery or somewhere you get to say, "Can I take your order? Please drive to the first window."

31. Be the last one to serve your plate; make sure the other members get their food first.

32. A friend of mine was a new chief years ago. On his first day, one of the departments in his county held a fundraiser. He spent time visiting and then helped with the dishes. If a chief can do dishes, you can, too. After a meal, get your hands in the sink to do the dishes, and be first to take out the garbage and mop the kitchen floor.

33. Don't tell jokes until you're accepted. If you're not sure, say nothing. Shut up.

34. Don't gossip. You'll be tempted, but seriously, don't. Just shut up. Gossip helps no one. Shut up. It's Golden Rule time for you.

35. Watch your temper. Chop busting is part of firehouse life. And if they try to get to you, watch yourself and your response. Odds are, you should have no response. Smile and have fun.

36. Help others with their assignments when you finish yours; that includes cleaning and training. Ask another probie to quiz you, drill with you, or practice a skill together. If they think that's BS, find someone else to work with.

37. Volunteer for assignments, stand-bys, special details, conferences, fundraising, teams, and special trainings.

38. Until you're tested as an expert or assigned to do something, do nothing at a scene until the officer tells you what to do.

39. Never turn your back on a rig that's backing up, and never get in the way of traffic. Stay out of the roadway on runs unless it's blocked, and even then, be *very* cautious. Don't trust the public. They don't see you. Watch the traffic.

40. A good analogy for your probationary period is that the department is loaning you the *temporary* title of "probationary firefighter." At the end of the specified time frame, your officers (with input from the crew) will decide if you get to be called a firefighter—no kidding. If you've proven yourself "worthy," followed some of this advice, and gotten along with your crew, their decision is easy. If not, their decision is also easy.

41. Watch what you eat. Stay fit. You don't want to puke in your mask at a fire.

42. Work out *every* day. You'll thank me when you have to climb six stories with 100-plus pounds of gear and then attack a fire.

43. EMS can be a pain, especially when abused by those who don't need EMS . . . but that's part of what we do. In most cases, EMS means

someone is having a really bad day, and you have a chance to help change that. Sometimes you'll win, sometimes you won't—just do your best. How would you want to be treated if you were in the patient's situation? What if it was your mom? EMS is a part of being a firefighter these days and it matters. More lives are saved by firefighters doing EMS-related duties than anything else.

44. When you learn the job of an engine company firefighter, remember that getting water on the fire can do more to save lives than almost anything else. When the fire is under control, the scene gets better: The smoke slows down, the fire stops burning, and so on. Learn the role of an engine firefighter . . . like an expert.

45. When you learn the job of a truck (ladder) company firefighter, your job is usually to vent, enter, search, and rescue people. Sometimes you won't have a waterline to protect you, so learn what to do about that . . . like an expert.

46. Study building construction. The building is your enemy, and you must know the enemy. Frank Brannigan said that. Who is he? Figure it out!

47. Learn to calm down. No matter how bad it is—and you are going to see some *really* bad things—relax. Don't get excited. We're supposed to *fix* the problem for which someone dialed 9-1-1. Control yourself and *think* about your responsibility on that run.

48. At a fire, have *no* exposed skin (getting burned sucks), use all of your protective gear, and don't breathe smoke. Smoke is a deadly poison that can get you either now or later. Look at what's burning: It's all man-made and, in most cases, made of fuels (plastics, etc.) that give off cyanide gases. Don't breathe that crap. Use your SCBA—that's what it's there for.

49. If you're lucky and work your tail off, you might earn the right to become a firefighter. Not long after that, you may have the privilege to learn how to drive the apparatus; when you do, don't drive like a lunatic. They *let you* drive the apparatus; it isn't *your* apparatus. Don't hurt your firefighters or the public. Drive sane so you can get to the fire.

50. You will have an elated feeling when you get on the apparatus and respond on your runs! Always wear your seatbelt, smile, and remember how great this job is. And in your first couple months, when you get back to the firehouse and no one is looking, call someone who knows what you're doing and tell them about the run.

I could go on for hours, but I think these first 50 tips will give you something to think about. Now go study and train. And shut up!

I love you Sonny Boy. *Do good.*

Love,

Dad